W9-DBH-113

HOW TO SPEAK WHALE

HOW TO SPEAK WHALE

A VOYAGE INTO THE FUTURE OF ANIMAL COMMUNICATION

TOM MUSTILL

NEW YORK BOSTON

Grand Central Publishing
Hachette Book Group
1290 Avenue of the Americas, New York, NY 10104
grandcentralpublishing.com
twitter.com/grandcentralpub

First U.S. Edition: September 2022

Grand Central Publishing is a division of Hachette Book Group, Inc.
The Grand Central Publishing name and logo is a trademark of Hachette Book Group, Inc.

The publisher is not responsible for websites (or their content) that are not owned by the publisher.

The Hachette Speakers Bureau provides a wide range of authors for speaking events. To find out more, go to www.hachettespeakersbureau.com or call (866) 376-6591.

Additional copyright/credits information is on page 267.

Library of Congress Cataloging-in-Publication Data

Names: Mustill, Tom, author.
Title: How to speak whale : a voyage into the future of animal communication / Tom Mustill.
Description: First U.S. edition. | New York : Grand Central Publishing, 2022. | Includes bibliographical references. | Summary: "What if animals and humans could speak to one another? Tom Mustill-the nature documentarian who went viral when a thirty ton humpback whale breached onto his kayak-asks this question in his thrilling investigation into whale science and animal communication. "When a whale is in the water, it is like an iceberg: you only see a fraction of it and have no conception of its size." On September 12, 2015, Tom Mustill was paddling in a two-person kayak with a friend, just off the coast of California. It was cold, but idyllic-until a humpback whale breached, landing on top of them, releasing the energy equivalent of forty hand grenades. He was certain he was about to die, but both he and his friend survived miraculously unscathed. In the interviews that followed the incident, Mustill was left with one question: What could this astonishing encounter teach us? Drawing from his experience as a naturalist and wildlife filmmaker, Mustill started investigating human-whale interactions around the world. When he met two tech entrepreneurs, who told him they wanted to use artificial intelligence (AI) to decode animal communication, Mustill embarked on a journey where big data meets big beasts, using animal eavesdropping technologies to train AI-originally designed to translate human languages-to discover patterns in the conversations of animals. There is a revolution taking place in biology, as the technologies we've developed to explore our own languages are turned to nature. From seventeenth-century Dutch inventors, to the whaling industry of the nineteenth century, to the cutting edge of Silicon Valley, How to Speak Whale looks at how scientists and start-ups around the world are decoding animal languages. Whales, with their giant mammalian brains, offer one of the most realistic opportunities for this to happen. But what would the consequences of such human-animal interaction be? We're about to find out"— Provided by publisher.
Identifiers: LCCN 2022019457 | ISBN 9781538739112 (hardcover) | ISBN 9781538739136 (ebook)
Subjects: LCSH: Animal communication.
Classification: LCC QL776 .M88 2022 | DDC 591.59—dc23/eng/20220524
LC record available at https://lccn.loc.gov/2022019457

ISBNs: 978-1-5387-3911-2 (hardcover), 978-1-5387-3913-6 (ebook)

Printed in the United States of America

LSC-C

Printing 1, 2022

For Dad, I wish you could have seen it.

And to Humpback Whale CRC-12564,
for setting me on this journey.

How do you expect to communicate with the ocean,
when you can't even understand one another?

—*Stanisław Lem,* Solaris

An illustration by the artist Sarah A. King of the view I had as the whale bore down on Charlotte and me.

CONTENTS

INTRODUCTION
Van Leeuwenhoek Decides to Look xiii
(NEW TOOLS PLUS INQUISITIVE HUMANS EQUALS UNEXPECTED DISCOVERY)

1 *Enter, Pursued by Whale* 1
(THE TWENTY-FIRST-CENTURY REVOLUTION IN CETACEAN BIOLOGY, AND HOW I JOINED IT)

2 *A Song in the Ocean* 19
(HOW WE SAVED THE WHALES BY DECODING THEM)

3 *The Law of the Tongue* 35
(DIFFERENT SPECIES ALREADY COMMUNICATE)

4 *The Joy of Whales* 53
(DO CETACEANS HAVE THE TOOLS TO TALK AND LISTEN?)

5 *"Some Sort of Stupid, Big Fish"* 71
(WHAT CAN WHALE BRAINS TELL US ABOUT THEIR MINDS?)

6 *The Search for Animal Language* 87
(LET'S AVOID THE WORD "LANGUAGE")

7 *Deep Minds: Cetacean Culture Club* 110
(HOW DOLPHIN BEHAVIORS SUGGEST THEY'RE WORTH TRYING TO CHAT TO)

8 *The Sea Has Ears* 128
(ROBOTS CAN RECORD WHALE COMMUNICATIONS WE NEVER COULD BEFORE)

9 *Animalgorithms* 145
(HOW WE CAN TRAIN MACHINES TO FIND PATTERNS IN CETACEAN COMMUNICATIONS)

10 *Machines of Loving Grace* 167
(GOOGLE TRANSLATE FOR WHALES)

11 *Anthropodenial* 194
(HUMANS UNDERESTIMATE OTHER ANIMALS… AND WHY IT MATTERS)

12 *Dances with Whales* 212
(IT'S TIME TO FIND OUT IF WE CAN SPEAK WHALE)

Acknowledgments 229

Notes 233

Photo Credits 267

Index 271

About the Author 285

INTRODUCTION

Van Leeuwenhoek Decides to Look

What if I had never seen this before?
—*Rachel Carson,* The Sense of Wonder

In the mid-seventeenth century, in Delft in the Dutch Republic, there lived an unusual man called Antonie van Leeuwenhoek. This is him:

Antonie van Leeuwenhoek holding one of his magnifying inventions in 1686, by Jan Verkolje.

Van Leeuwenhoek was a businessman, a draper. He was also a high-tech inventor. The previous fifty years in Europe had seen the rapid development of tools for magnification—telescopes and microscopes. Most worked on similar principles, with two glass lenses fitted into a tube. Looking through these lenses gave the user superhuman powers, bringing distant planets and tiny objects into better view. They were also very rare: Few people had learned how to grind, polish, and fit the glasses, and many guarded their secrets closely. For van Leeuwenhoek, as an apprentice draper, "microscopes" (from the Greek words for "small" and "to look at") were also the tools of his trade, useful for inspecting the quality of the fabrics he bought and sold. The early microscopes could magnify up to nine times, while later iterations could zoom in farther. But the multilens design had a flaw—the more they magnified, the more distorted the lenses made the image, and above about twenty times magnification it was hard to make anything out.

In Delft, van Leeuwenhoek had been secretly pioneering a different technique. Instead of using a series of lenses, he became expert at crafting tiny individual spheres of glass, some just over a millimeter in diameter, which he mounted onto folding metal brackets. By placing an object onto the bracket, holding the glass sphere very close to his eye, and looking through it into a source of light, he found he could magnify his subject by up to 275 times and with little distortion. He is thought to have made more than five hundred microscopes during his lifetime. Recent studies have found the focusing capabilities and clarity of his devices comparable to those of modern light microscopes.

Van Leeuwenhoek did not just use his revolutionary magnifying technology to inspect the weave of the cloth he sold. He explored the world beyond his trade. While other microscopists had enlarged and explored the visible—things like insects, or cork—van Leeuwenhoek discovered entire invisible realms. In a thimbleful of water from a local lake, empty to the naked eye, he was astonished to spy hordes of "animalcules"—tiny animals, bacteria, and single-celled organisms. Everywhere he looked, he found scurrying swarms of previously unknown creatures: in the world around us—in rainwater and well water, and within our bodies—samples scraped from his mouth and taken from intestines. Van Leeuwenhoek was entranced, writing that "no more pleasant sight has yet met my eye than this of so many thousands of living creatures in one small drop of water, all huddling and moving."

At the time, people were unable to see the eggs of fleas, eels, or mussels, so they assumed they didn't exist. Rather than growing from eggs as larger animals did, it was believed that these small animals came into being through a process called "spontaneous generation," whereby fleas could spring into life from dust, mussels from sand, and eels from dew. Van Leeuwenhoek's tools revealed the previously imperceptible eggs of these animals, and doomed this theory. He was obsessed with the new world he had discovered: red blood cells, bacteria, the structure of salt, the muscle cells of whale meat. He investigated the still-mysterious world of human reproduction, perceiving within semen tiny moving bodies with tails—sperm. When I think of this moment, I wonder both how astonishing this must have been, and whose semen he got hold of.

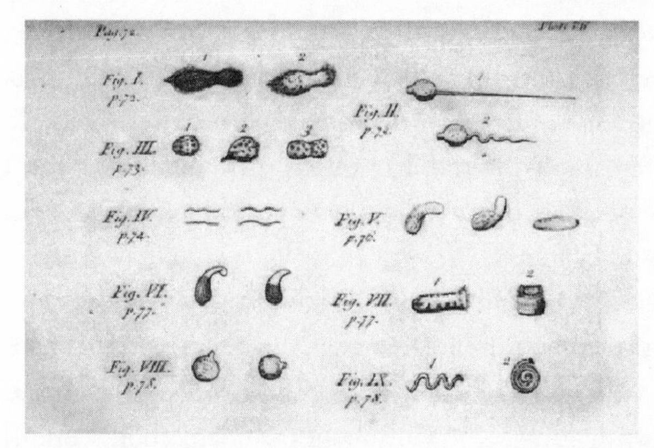

A copy of the illustrations made by van Leeuwenhoek of the animalcules he had discovered. Figure IV is thought to be the first printed rendering of a bacterium.

Across the Channel, in England, the natural philosopher Robert Hooke had himself been experimenting with microscopes, adding to and modifying their lenses and exploring the structures of snowflakes and the hairs of fleas. The drawings he published of these hidden worlds caused a public sensation. The diarist Samuel Pepys stayed up until 2 a.m. reading Hooke's book in bed. Poring over the fold-out illustrations, he wrote that it was "the most ingenious book that ever I read in my life." Van Leeuwenhoek wrote to Hooke and the other learned experimenters of the Royal Society (then called the Royal Society of London for the Improvement of Natural Knowledge) and reported his findings. At first, many did not believe the "exceedingly curious and industrius" merchant, despite credible witnesses. How could there be whole domains of life completely invisible to us? Van Leeuwenhoek complained that he "oft-times hear it said that I do but tell fairy tales about the little animals." It didn't help that he closely guarded his microscopes and his methods for crafting them.

In London, Hooke worked to reproduce van Leeuwenhoek's results. It took many attempts to replicate the exquisite, tiny glass spheres, but when he finally succeeded on November 15, 1677, he gazed into rainwater and beheld tiny moving creatures; "surprized at this so wonderful a spectacle" he, too, "verily believed" them to be animals. Seeing was believing. Van Leeuwenhoek was duly made a fellow of the Society and is today widely acknowledged as the father of microbiology. His inventions allowed us to view the microscopic life that has always surrounded us—but just as crucially, he possessed a mind curious enough to look where others assumed nothing would be found.

A few centuries on and our culture has changed. When someone sneezes in the street, you picture the germs spraying across you. When you worry that your mole looks a bit funny, you conjure up visions of tiny cancerous cells furiously dividing. Knowing about the microscopic world changes our lives: We wash our hands and wounds, we create and freeze embryos. We know that hidden within each of our bodies there are as many bacteria as human cells. An invisible ecosystem. His decision to look has transformed our behaviors, our cultures, and how we see ourselves.

This is the legacy of van Leeuwenhoek's invention. We cannot unsee what he first spied.

* * *

What further invisible worlds might we now discover? You are already part of one new frontier. Since the seventeenth century our tools for looking have proliferated, and many are now pointed toward ourselves: Security cameras track you walking down the street, the thermometer and gyroscope in your iPhone senses you shifting in your sleep as the room cools. So much is now tracked: When you sleep and when you dream. Where you live and where you go. Your fingerprints, voiceprints, iris pattern, gait, weight, ovulation, body temperature, likely infections, breast scans, the steps you take, the shape of your face and the expressions it can pull. What you like, what you don't. Who you like, who you don't. The songs and colors and objects you are drawn to. What turns you on. What you think is funny. Your name and your avatars and handles. The words you use, the accent with which you speak. And we're just getting started. You are now remembered not just by your friends and family, but by computers you have never met—what they sense of you is crystallized in *data* and transmitted through the internet to vast servers, where it sits with the data of billions of other humans. Your data are accumulating faster than any memoir you could write, and when you die, they will outlast you. And within these data, other machines are trained to find invisible patterns.

For the past couple of decades, many of our brightest engineers, mathematicians, psychologists, computer scientists, and anthropologists have been scooped from universities to work for Alphabet, Meta, Baidu, Tencent, the other giant information corporations, as well as the governments of the United States and China. In the 1940s, these minds might have been set to work on splitting the atom at the Manhattan Project; in the 1960s, they might have been employed designing spacecraft at the Jet Propulsion Laboratory. Today, clever young people are richly rewarded for finding new ways to record, amass, and analyze human data. Using invisible patterns in language, their machines can translate between human tongues without ever being taught how to speak one; using hidden patterns in faces, they can tell when a human smile is genuine better than a human can. We are begrudgingly accepting this accumulation of our data, as well as the fact that we can be manipulated by those who understand these patterns within it.

In all this, it is easy to forget that we are animals, human animals. All these patterns—our bodies, behaviors, and communications—are *biology*. The tools we have made for finding invisible patterns in humans can work on other species, too. Like van Leeuwenhoek's microscopes—useful for evaluating cloth, but also good for discovering the origins of fleas—many of our tracking devices, sensors, and pattern recognition machines were originally developed to sell things to people effectively but are now being turned outward, toward other species and the rest of nature. And they are, in the process, revolutionizing biology.

This book is about some of the pioneers in this new age of discovery: the decryption of the natural world. It is a journey to the frontiers where big data meets big beasts, where silicon-based intelligences are finding patterns in carbon-based life. It focuses on some of the most mysterious and fascinating animals—whales and dolphins—and how recent technology has radically changed what we know about their hidden lives and capabilities. It explores the way underwater robots, massive data sets, artificial intelligence (AI), and changes in human culture are combining to transform how biologists decode cetacean communications.

This book is about learning to speak whale. About whether, with all that is changing in our science, technology, and culture, such a thing could ever be possible. As we turn our pattern-finding machines away from ourselves and focus them on the utterances of other species, I have come to wonder if we will be changed by what we find, much as the microscopic worlds that van Leeuwenhoek saw through his glass spheres changed us. Could our discoveries compel us to protect these animals?

I know this all sounds a bit far-fetched. I thought so, too. But I didn't just come up with this story: It found me, and I stumbled after it. It began in 2015 when a thirty-ton humpback whale leapt out of the sea and landed on top of me.

1

Enter, Pursued by Whale

They say the sea is cold,
but the sea contains the hottest blood of all.
—*Captain James T. Kirk,* Star Trek IV: The Voyage Home

On September 12, 2015, I was kayaking with my friend Charlotte Kinloch in Monterey Bay, off the coast of California. We'd left the shore around 6 a.m., along with a guide and a half-dozen other kayakers from Moss Landing, a deepwater harbor halfway up the long bay between the coastal cities of Monterey and Santa Cruz. We were split into pairs, and each pair was allotted a two-person kayak. It was cold, misty, and so still I could hear the water dripping from our paddles onto the sea's surface. Within the calm of the harbor walls, sea otters floated on their backs at rest, eyeing us from a distance, clamped to one another in fluffy rafts. As we turned past the piled boulders of the breakwater into the open ocean, groups of sea lions rolled about on the surface all around us, like the tops of turning cogs in a giant underwater machine, whiskery and snorting. The fog around us diffused the morning light so that it felt like paddling in a lightbox, and often little could be seen, but life was all around us. Above, pelicans cruised to the jagged cries of gulls.

I peered at the gray, almost metallic sea. Beneath us now was an underwater chasm deeper than the Grand Canyon. Although we were within earshot of land, it was already hundreds of fathoms deep, a great crack

Beneath this splash are Charlotte and me, our kayak, and a humpback whale.

stretching thirty miles out from the coast far out to the sea. A freak of geology—the third largest underwater valley in the world—it channeled deep, food-rich seawater up to the surface, where the maritime alchemy of sunlight and nutrients fed an astonishing food chain considered a wonder of the natural world. In the 276 miles of shoreline and six thousand square miles of ocean, the National Marine Sanctuary contains such an abundance and diversity of life that it is known as the Blue Serengeti. On land there are just a few places, like its namesake the Serengeti in Tanzania, where you can witness megafauna. On most continents the largest animal you are likely to see is a cow. Yet many giant beasts remain in the seas. Mostly they dwell far from human eyes, in polar waters or clustering around remote island chains. But here, because of the canyon, mingled the largest aquatic creatures on the planet: great white sharks, leatherback turtles, giant ocean sunfish, elephant seals, humpback whales, killer whales, and the vastest of all the megafauna, blue whales. Right next to the shore, alongside one of the biggest sprawls of humankind, just down the road from San Francisco and Silicon Valley.

Our guide, Sean, was a young, bearded, brown-haired bloke who looked like he spent more time with a kayak-skirt hanging from his waist than dressed in land clothes. Sean had explained that if we saw any whales, we should keep a hundred yards away from them. They were wild animals and it was on us to stay out of their way, not the other way around. There were many kinds of cetaceans in these waters. Gray whale mothers escorting their young calves along the coast from their birth waters in Mexico; killer whales lurking to hunt them; fin whales, minke whales, cruising through, chasing plankton swarms; and Risso's dolphins hunting squid.

As we paddled out of the harbor, only a few minutes passed before we saw the whales. They were everywhere. The morning fog lifted to reveal their spouts shooting up from the surface in all directions; whale breath marked the air along the sandy coast toward Monterey and out to sea. As a conservation biologist and wildlife filmmaker, I have been fortunate to see a lot of whales of many different kinds. But I had never experienced anything like this. There were so many. At first, all those we spotted were distant, half a mile off. Then a group of three popped up a stone's throw away, one after another, moving quickly. Before long, more appeared and disappeared behind us. Sean told us to stick close together, and we backpedaled to keep our distance. With no wind or waves, the sudden, plosive exhale of a whale surfacing felt scarily loud and close, something between a whinnying horse and a gas canister being depressurized. Their breath, like stale, fishy broccoli, carried downwind toward us.

Often sighting a whale can be anticlimactic: Mostly you see them when they come up to breathe, and from high up on the deck of a boat it can be like fleetingly glimpsing a big exhaling log. It's often hard to get a sense of their scale. Yet, from the kayak, it was remarkably different. As we watched them along the water level, we felt their size, and their power.

The whales we were seeking that morning were a species called humpback whales (*Megaptera novaeangliae*), one of the largest of all the cetaceans—the name given to the group of mammals that includes whales, dolphins, and porpoises. When a humpback whale is born, it already weighs as much as a white rhino. The adults swimming around us were each mostly the size of an airport shuttle bus. The flat wash of the misty light brought out all the detail of their skin, which resembled the texture of a cucumber with a filigree of cracks and scars, the muscular ridges along their double nostrils, clamped shut on the tops of their heads. They were blue-gray on top and paler underneath, with long, armlike pectoral fins.

We'd been told that the whales were feeding on a school of fish that stretched over a mile underwater, and there was clearly a great whale feast happening beneath us. Humpbacks are binge eaters, hunting and swallowing hundreds of fish at a time. They are migratory: In the summers, they move to cooler waters, such as Antarctica, Alaska, and Monterey Bay, where they will spend most of each day eating. They pile on the pounds, feasting month after month. Then, in the winters, they fast, not eating for months. They swim to warmer tropical seas where they woo one another, shed their parasites, and birth their young (called calves). Humpbacks are unusually "surface-active" whales; they often lift parts of their bodies out of the water, or roll around on the surface. When lunging at their prey, they can suddenly shoot most of their heads out of the water, mouths agape. When they dive, they fold gracefully and their tail flukes can extend well clear of the surface. In the tropics, they mainly rest and move little, conserving strength for the long journey back. The peace is sometimes broken by the males (called bulls), who charge around after the females (cows) in "heat runs," battling and jostling one another in bloody, dangerous competitions. Their annual migrations are the longest of any mammal, spanning entire oceans. By the time they return to their feeding grounds, their fatty blubber is so depleted the outlines of their spines are clearly visible. So humpbacks don't mess around in Monterey. When it's feeding time, they gorge.

All around us the humpbacks were moving, and fast. They seemed to be clustered in small groups of three or four and turning often. I've learned since that these whales can work in teams, using their bodies and walls of exhaled bubbles to trap balls of fish and push them toward the surface before lunging at them together. In these maneuvers, different whales seem to take on different roles. Unusual for mammals that cooperate, the teams of whales often aren't related to one another, and they stick together year after year, traveling across thousands of miles in convoy. I watched a group of four whales surface, their bodies aligned, pectoral fins overlappingly close. In unison, they exhaled, inhaled, and immediately disappeared. They seemed like volleyball players fist-bumping between points.

These relationships have been called friendships (though scientists generally refer to them as "stable multiyear associations"). Bobbing in our kayaks, toes numb, mouths agape, we watched them feed. I was later told that at least 120 whales were identified in the bay that day. Sometimes they would slap their fins on the water with a great *phat* noise ("pec-slapping"), or even raise their heads above the surface to the level where their eyes could look about them in the air, a behavior known as "spy hopping." Toward the horizon we saw a couple of partial breaches—whales throwing themselves up out of the water and crashing back down in a white explosion with a *whump* noise, like distant thunder. I didn't realize at the time that this was, even for Monterey Bay, an unprecedented feeding frenzy. We'd chanced upon the greatest concentration of whales, with the calmest weather, the closest to shore in living memory.

I looked over at Sean, our guide, and noticed that he did not seem relaxed. His eyes flicked back and forth over the four boats in our little flotilla; he regularly called out to us to come back together if we drifted too far apart and to paddle backward as new whales appeared. Of course, whales can move a lot faster than kayaks. As the morning wore on, three or four whale-watching boats and other kayakers joined us. We were so close to the beach that a stand-up paddleboarder had even made it out. I had long since stopped caring about how cold and damp it was, or the fact that all feeling from my bottom was gone. After a couple of hours, Charlotte—who had before that day never seen a whale in her life—and I turned our craft away from the whales, and with the rest of our group, we headed back to shore, hyperattenuated and awed.

We had made it about halfway to the harbor when suddenly about thirty feet in front of us an adult humpback erupted from the sea, shooting impossibly upward—as if a building had grown out of the ocean, as Charlotte would later describe it. When a whale is in the water, it is like an iceberg: You see only a fraction of it and have no true conception of its size. Each foot of humpback whale weighs about a ton, and adults range from about thirty to fifty feet long. An animal three times the weight of a double-decker bus. Can you imagine what that looks like hovering above you? One moment we were on the flat, calm sea going home, the next this gargantuan, living mass of muscle and blood and bone was in the air, arcing toward us. I remember noticing the grooves on its throat. *Ventral pleats*, I thought to myself. And the next thing I recall is being underwater.

A humpback whale is three times bigger than the biggest *T. rex*; its sixteen-foot pectoral fin is the largest and most powerful arm in the history of life on Earth. If you were to X-ray a humpback's pec fin you would see your own arm, writ monstrous: shoulder blades into humerus bone, joining radius and ulna, hand bones and fingers—a legacy of their lives on land before their ancestors returned to the seas. As the whale came down onto us, the force of the impact punched the kayak beneath the water, and we were sucked down with the sinking whale, leaving just an explosion of spray where we'd been a moment before. Underwater, dislodged from the kayak, I spun, a toy person, tumbling in the freezing water faster than I thought possible, my stomach yawing with the feeling you get when you jump off something high. My eyes were open but I could see nothing but whiteness. I sensed that the whale was still very close. And then I felt it move away without touching me. The white of the explosion turned to dark seawater. It was only at that point that I felt fear. Until then, it had been a matter of facts: There was a whale above my head and I was going to die. Some reptile part of my brain now rationalized that the only reason I wasn't yet dead must be that I was in shock and couldn't feel that my body was broken into pieces. Soon, I'd surely be hit by the pain and lose consciousness. But miraculously I felt my lifejacket tug upward, and I kicked with it toward the light.

I was certain that Charlotte was dead. As I broke the surface and looked around, I saw her head. Her living head, attached to the rest of her body, eyes wide open and mouth pulled tight in a grin of adrenaline and fear. I felt sheer delight. We were alive.

How the fuck were we alive?

We swam over to our kayak, which was filled with water and lolling on the surface, and clung to it. Its nose was dented and deformed from the impact, and there were scratch marks from where the barnacles living on the whale's skin had scraped it. Later, I wondered how much force it must take to dent the rigid molded plastic of a kayak floating on water. If I punched a rubber duck floating in a bath as hard as I could, it wouldn't leave a mark. Scientists have estimated the forces involved. To breach, a humpback must reach speeds of up to twenty-six feet per second, an astounding velocity for a thing the size of a truck moving through water. For a large adult whale to breach like this, they estimated it would take a release of energy equivalent to about forty hand grenades. It felt as though we'd survived a lightning strike.

Other kayakers paddled over, seemingly more upset than us—which is understandable considering they'd thought they'd just watched us die. As someone fished Charlotte's flip-flops from the water, a whale-watching boat chugged up alongside us. We looked up at ranks of tourists leaning over. Some shouted to ask if we were OK, while others recorded us on their phones. Most had been looking the other way, out to sea. They assumed we'd been knocked out of the kayak by the splash, not that a whale had actually hit us. We dangled off someone else's kayak in a state of euphoria and shock, while someone turned ours over to empty it. We were safe. Just then, a whale began moving toward us along the surface of the water. "He's back for more!" joked a nearby kayaker.

I laughed, but I was unnerved. While I knew that these whales don't eat people and in fact cannot, having no teeth and a throat the width of a grapefruit, I was also aware that they don't often breach on people, either. Just as the approaching whale's head felt like it would hit us, it tilted its front end down and dove. As humpbacks bend into the dive their backs arch distinctively, showing prominently the bulge ahead of their dorsal fin that gives them their name. As the long spine curved and the whale's head plunged toward the seabed, other parts of it were still moving upward. Like carriages on a train, sections of the whale first rose, then disappeared beneath us: the dorsal fin, then the thick, stocky caudal peduncle—like the tail of a diplodocus, which narrows to a human torso's width—before the great tail fluke finally emerged, glistening, into the air, trails of water dripping off the tips of each half of the massive paddle.

Floating in the water, I was transfixed by the sight so close in front of us, beloved of whale watchers. The huge, black, heart-shaped flukes shone in the gray light, the tail's end alone the size of a horse. *This is called fluking,* I thought. *It's doing that so the weight of its tail helps it overcome its buoyancy, allowing it to sink.* Where it dove, it left a mark like a big pancake on the water. A whale footprint. If I had stretched my feet out beneath me, I think I could have touched its body as it passed under. Instead, I wrapped my feeble, stubby land legs around the kayak I clung to like a sloth. I then remembered that a whale had just jumped on us, and we had survived. I turned and said this to Charlotte. In more colorful language, she said she was aware of this but needed me to just shut up until we got back to land.

Eventually, the whale watchers went back to whale-watching, and we climbed back into our now-drained kayak. Sean, clearly distressed, tied a rope between his kayak and ours and struck for the harbor. The two giant chimneys of the disused power station behind Moss Landing loomed out of the fog, which had lifted into a thin layer. We were shivering now. On our way, we passed schoolchildren and their teachers heading out. Everyone was in high spirits. "A whale just landed on us," I said as we passed a group of them, but they just grinned at the odd, wet British man and continued out to sea. Back at the base they gave us each a Monterey Bay Kayaks baseball cap and some hot chocolate. Nobody said much to us.

It all felt strangely awkward, like there had been a faux pas. I'm not sure any of us could quite compute the power of what had happened, and the awfulness of what had nearly happened. Maybe they were worried we'd sue. (I later learned that they stopped leading whale kayak tours; the word was that their insurance would no longer cover it.) A friend drove us back to an Airbnb we were renting and, en route, Charlotte finally burst into tears. When I leaned forward in the car to tie my shoelace, my nose leaked streams of seawater—which had been trapped in my sinuses from our underwater cartwheels. All I could think about was the beautiful violence that I had been briefly part of, and how no one would believe us. It occurred to me then that I'd left my two GoPro cameras in the car that morning. Charlotte had suggested I bring them, but I'd decided not to—after all, everyone's whale video looks the same.

We rejoined our friends at our rented beach house. It was the end of a long group holiday together and they were all ready to leave for the airport. I was staying on to go camping with some other friends nearby.

"You're late," said our friend Louise. "We've had to pack your stuff for you and you've missed breakfast."

"A whale jumped on us," I said.

"That's fine," said Louise, "but if we don't leave soon, we'll be fined for late departure."

I hugged Charlotte, who had mostly stopped talking. She had tried to explain what happened to her husband, Tom, who was mainly disgruntled as he loves both whales and risky exploits. We ate some leftover breakfast. And then everybody left. Charlotte fainted on the flight home and had to be given oxygen.

Outside our beach house I sat by the road waiting for my friend Nico and his parents to pick me up. I realized that the only person who could corroborate my tall tale had left. I squeezed into the backseat of the car with Nico's mother and his then-girlfriend, Tanya. I told them what had happened, and while I think they believed me, Nico's mother seemed more curious to know what Tanya's parents did. But I couldn't just keep repeating the story, so we moved on to other topics. A few hours later, in the mountains of Big Sur, we reached the campsite up in the pine trees. As darkness fell, I looked out at the Pacific from the dusty hillside and drank a beer while a group of other campers nearby played music. We had no phone signal, so I was left to process what had happened by myself. I wondered whether anybody believed me. I mean, who would really believe a thirty-ton whale had breached onto us?

That night, lying awake in my tent, I looked up into the dark and saw the impossibly large body above me: the seawater streaming off it, knobby tubercules—lumps that house a whale's whiskers—scattered around its head, the barnacles on the edges of its fins. It had seemed so much more massive in the air than in the sea, but also like an absurd joke. There had been little time to be scared in the moment itself, but as I reflected on the experience, I felt my heart race. People have often asked me in the years since if I was traumatized, but I don't think I was. In all honesty, I was exhilarated. What a thing to look upon, to feel. I lay down and closed my eyes and burned the images of what had happened into my memory so that I could never forget it.

The next day we drove back to San Francisco, and as we left the National Park, cell phone reception returned. Tanya and Nico's mother got into an argument about pets; Nico smoothed it over. Crammed in alongside them, I trawled the internet searching for something—a photo, a blog, anything to prove it was real. And there it was. In a bizarre coincidence, at the moment the whale leapt out of the sea, a man named Larry Plants on a nearby whale-watching boat had been filming on his phone. You could see us paddling along, and then suddenly the whale emerged and crashed onto us, and Charlotte and I momentarily disappear in a white explosion, before bobbing up again six long seconds later. The video was complete with the eerie soundtrack of Larry shouting in triumph, "I got it, I got it on video," and a woman nearby screaming, "The kayak, the kayak!" He'd sent it to the whale-watching company, which then uploaded it to YouTube. It already had more than one hundred thousand views.

A frame grab from the video by Larry Plants.

Realizing that soon many more people would watch the video, I thought I should probably call my mother, Caroline. I told her that I had nearly been killed by a falling whale but that I was fine, and was coming home.

"Honestly, Tom," she said. And then: "What would your father have thought?" It had been my first thought, too. My father, Michael, loved strange beasts and stories of the sea. But we couldn't ask Dad because he'd died a few months earlier. I was at the stage of grief where I'd still think to phone him when something interesting happened before remembering, with that strange mix of shock and embarrassment, that he was gone.

While I was at the airport, *Good Morning America* called to interview me. By the time I landed in London the next day, the video had four million views and counting. Our encounter had gone viral and now had a digital life of its own. I took the tube from Heathrow and got out at Dalston Kingsland. It was a beautiful early autumn evening, the light low and golden. People were in the streets drinking and shouting as if nothing had changed. How could nothing have changed when two days ago a whale had been towering in the air above me? I remembered walking the same route and experiencing a similar sensation on my way back home from Dad's cottage, the day after he'd died. I'd looked at the other people, knowing that the world was not the same as it had been before, and yet there were all these people acting as if nothing had happened. More than six million people watched the video. It seemed to touch some grim fascination, the spectacular and unexpected collision of a giant, hidden, mysterious beast and two tiny humans.

Looking for meaning in our near miss was perhaps a fool's errand, like a squirrel searching for meaning when a truck thunders an inch from its nose on a country road. But a few days after the whale collision, my friend Professor Joy Reidenberg at the Icahn School of Medicine at Mount Sinai in New York wrote to me, saying that she had been thinking about the breach. A whale specialist whom I have worked with on many films, Joy has spent her entire life studying their anatomy. From her laboratory seventeen floors above Central Park, surrounded by the skulls of killer whales, and medical students dissecting human cadavers, she wrote that the breach the whale had executed seemed weird—that it started out going one way, and then appeared to change its course in the air above us. Instead of landing on us, it twisted and veered away, only clipping us with its fin. "I think you two survived because the whale cared about trying not to hit you," she wrote.

Charlotte and me, happy and alive.

Was she right? Did it really try to avoid us? It didn't crush us as it fell, or injure us in the water, and it moved away very slowly. New Age friends agreed with Dr. Reidenberg, believing it was a sign from the universe. However, other whale experts gave different opinions. Some said it was likely an aggressive act, that the whale wanted to hit us. Others believed that by breaching it was just displaying a behavior common after feeding, that it was signaling something to other whales.

It became apparent that part of the problem with trying to understand what the whale was doing when it breached onto us is that we don't know why they breach at all. I find this extraordinary. One of the biggest animals in the history of life on Earth can propel itself out of the element it lives in, and execute a spectacular special-effects ballet move, and we don't know why. Some think whales jump to rid themselves of the giant lice and barnacles that reside on their skin. Others that it is a display of strength, or that they do it as play, or practice. The most popular theories are that the breaches are somehow involved in communication: whales use vocalizations to interact, but the sea can be noisy, and the sounds produced in a breach are extremely loud and, traveling through water—a far better conductor of sound than air—they can be heard many miles away. Had we strayed into a whale conversation, accidentally punctuating a sentence of splashes? Breaches could mean all these things to whales, or none of them. In Joy's words, "No one really knows what breaches are about. No one really knows for sure what happened to you. It's like asking someone who's dancing down the street, 'Why are you dancing down the street?' Well, maybe you're happy or maybe you're crazy or maybe you've got an ant in your shoe. You can't get inside its head or ask a whale, 'Why did you do that thing?' "

And it was true, you couldn't ask a whale.

The news cycle took us and ran. We were the novelty story at the end of bulletins around the world; we were in every newspaper, featured in *Time* magazine, name-checked on Japanese game shows. Reports and misreports sped across the world. In one appearance on breakfast TV, the interviewer asked us, absurdly: "When the whale jumped onto you... did you realize it was a whale?" Yes, I told them, we did realize it was a whale. We became a meme, a gif, a shareable video token of epic failure. The *Sunday Times* cartoonist converted us into a sketch, with Charlotte as Prime Minister David Cameron and me as Chancellor George Osborne in a rowing boat, with the whale as Labour leader Jeremy Corbyn leaping onto us. I was grimly fascinated with rewatching the video of the crash. Replaying it in slow motion, I realized the figure at the back of the kayak shrinks as the whale falls. That was me, trying to flip the craft over. But the stick woman at the front stays the same, frozen and erect. Charlotte had been so much closer to the whale than me. She had stared up at it the whole time until there was nothing but spray. Our mothers were quoted in the news: Charlotte's said she was never allowed to go in the sea again, mine that she was relieved I was OK but she thought it would have been a fitting way for me to go. And then the news moved on and that was that.

But things weren't the same. I was the whale boy, a lightning conductor for whale fanatics, and everyone I met seemed to have a whale or dolphin story. A retired Navy submariner from Yorkshire told me about listening to whales singing around his ship as it moved through the deep, their calls ringing through the hull. He felt they were playing with the submarine. A scientist told me how a gray whale had approached her in a Mexican lagoon, raised its head, and lolled against the side of her small boat with its mouth open. She reached inside its mouth and rubbed its enormous quivering tongue, their eyes locked the whole time. A book publisher told me she'd been swimming with wild dolphins in Australia. One swam over and buzzed her, seeming to scan her body with the sonar-like echolocation organs in its head. It was intently interested in her and no one else. The guides told her the dolphin was pregnant. The publisher discovered a few days later that she, too, was pregnant.

I received lots of messages from children and went to schools to try and explain what had happened. One wrote to me: "Dear Tom . . . And How diD whale Jump on You?" before continuing, "and Do you have eny Firends?" Through the whale I had indeed made friends. Other people who had experienced strange interactions with cetaceans—and people hooked on them. I have loved whales since I was tiny. One of my earliest memories was the disappointment I felt on learning that Wales, the country, was not a giant dolphinarium. My family holiday albums are full of postcards of orcas, and my first teenage summer job was on a whale-watching boat. Until I ended up on whale YouTube, I had never thought of it as a resource—but after the incident, I went down cetacean wormholes. Soon enough, I'd spent so much time looking at videos of whales and dolphins that the algorithms of my internet browser caught on, and I was bombarded with banner ads for Antarctic whale cruises and aquaparks.

There was a video, shot at night, of a scuba diver filming manta rays in the murky water, lit by underwater torches. A bottlenose dolphin approaches him; in its pectoral fin is a large fishing hook, the attached nylon line wrapped around the dolphin's front half. The diver beckons to the circling beast, who swims straight to his hand and lies still in the water while the diver feels along the line and wiggles the hook. Over the next few minutes, with hands, knife, and scissors he works the fishing gear from its fin, his fingers running up its length to the wild animal's mouth and back as he prizes the tangle from the flesh. Could the dolphin really have been asking for help? Could it really have been offering its fin to the human? My biologist's training had instilled an anthropomorphism alert for these sorts of thoughts. But how else could it be explained?

In another video, a scientist is filming two humpback whales underwater when one of them rolls her onto its back and pushes her along with its fin. Eventually, when she escapes to her boat, her colleagues scream that there had been a tiger shark in the water and that the whale had protected her. "I love you, thank you!" she shouts at the whale as it lingers near their vessel. A kayaker in Canada sings to a group of bright white beluga whales—and to his astonishment, one of the belugas mimics him and sings back. He slips out of his kayak and swims in the green water, singing gargly human underwater songs, the beluga chirruping and squeaking along as it eyes him, swimming in parallel. What did this trans-species aquatic duet mean, if anything? It struck me that before iPhones and GoPros, these videos would be anecdotes, retold and lucky to be half-believed. But as videos they could not be as easily dismissed. Of course, I could hear the scientist in me say there was selection bias: that videos of dolphins ignoring divers, of belugas turning mutely away from kayakers, and of humpback whales leaving swimmers vulnerable to sharks don't go viral. Nevertheless, as I devoured each new human-cetacean interaction, I wondered what it all added up to. Were there interspecies insights in these digitized encounters? Even in London, where I live, whales seemed to be in the news all the time. A beluga whale came to live in the Thames, a gleaming white Arctic sea beast longer than a horse, swimming downstream of a city of seven million people, most of whom had never heard of such a creature. It swam up the estuary and buzzed and clicked for weeks, before disappearing again.

These were incidents that demanded explanations and answers, but in many cases, what was transpiring already seemed intuitively obvious. The whales and dolphins were interacting, perhaps even communicating, with people. What, if anything at all, was the humpback that leapt onto us trying to say? But then my anthropomorphism alarm went off again and I felt silly. Still, the question remained. I couldn't quite shake it.

I'd been making wildlife films for a decade by this point, specializing in conservation and stories where people and nature meet. I secured a commission to make a film for the BBC Natural History Unit and the American broadcaster PBS about the community around Monterey Bay, and how their lives interacted with the whales there. I set about finding out as much as I could about what had happened to me, seeking to link this into the wider story of what was happening to humpback whales in the Pacific off California, as well as human-whale interactions globally. I spent months on boats with scientists, obsessed whale watchers, rescue teams, and fishermen. Whale books and research papers piled up in my head, alongside all those weird YouTube videos. Our little crew of three would head out at dawn to film the whales each day, getting out onto the water before the afternoon winds roughed up the sea and made keeping shots steady impossible. When seas were rough or whales hard to come by, we slept on piles of lifejackets. Peering down the viewfinder, anticipating where a whale might pop up, moving the focal point of the camera along their bodies, trying to find ways to frame their colossal shapes, playing back their exhalations and slaps in super-slow motion, amplifying their sounds and repeating them on loops, catching glimpses of their eyes sometimes staring straight at us—the more time I spent near them, the more enigmatic these creatures became. What is it like to be a whale? Did thoughts and feelings anything like our own stir inside them? At the end of each day, wind-burned and exhausted, I closed my eyes and still saw the sea, my internal gyroscope reeling from hours staring down a camera tube at the shifting horizon.

Author films a tail-slapping humpback in Monterey Bay.

Over the course of the filmmaking process, three things happened. The first was that every week seemed to herald a new revelation in the world of cetaceans. New populations of whales were being found, new behaviors witnessed, even new species discovered. Imagine if new species of elephants were being identified with such regularity. These animals can be up to twenty times the size of an elephant, yet in the last few years scientists have discovered what is highly likely to be a new species of mammal-hunting orca in Antarctica, a mysterious new deep-sea whale from New Zealand called Ramari's beaked whale, and a massive new filter-feeding whale species called Rice's whales in the Gulf of Mexico. In the Indian Ocean, the recordings of atom-bomb-detecting underwater microphones were used to discover two new populations of pygmy blue whale, told apart by their distinct songs. And within each new group there were new patterns of behaviors, new forms of communication, and, many scientists believe, new cultures. Of course, these are only new to us; the whales themselves have been there since before our time.

The second thing that happened during the course of filming was that I was surrounded by machines, and not just my own cameras. Drones flew overhead filming and measuring whales, research vessels dangled directional hydrophones, even the seabed was rigged with cameras, microphones, and other sensors. Machines were used to analyze the whales' feces, DNA, and mucus. There were scientists piloting remotely operated vehicles with robotic arms and probes; six-foot-long missile-shaped vessels piloted themselves under the waves far out to sea for months at a time. Scientists placed recorders on the whales, tracking their movements and documenting their world from their perspective while satellites followed them from space. On top of this, thousands of tourists took to the water every day, filming and photographing every cetacean they saw; each whale-watching boat had its own drone pilots, cameras on poles stuck over the sides, their own digital photo databases. Whales were being recorded more completely, more intimately, and more consistently than they ever had been before. These hugely innovative tools were changing what was possible in our relationship with the natural world—and none had existed when I had done my biology degree at the turn of the century. Like the period of microscopic discovery ushered in by Antonie van Leeuwenhoek's magnifying devices, this golden age of whale biology was driven by technology...and curiosity.

The third thing I realized while filming was that other powerful new machines were being used to make sense of all this information. And this fact had personal resonance. Using new databases of humpback whale photographs, which were set up just a few weeks before our near death, local citizen scientists were able to identify the whale that leapt onto Charlotte and me. They named it "Prime Suspect." This discovery was made by a computer algorithm specifically invented for the purpose of finding patterns in photos of humpback whales. Another algorithm was loosed on years of audio recordings from the seafloor, helping to reveal that whales in Monterey Bay were singing—a surprise to many who thought whales sang only in their winter tropical homes. It turns out they were singing day and night, all through the winter. This was a vivid example of a profound and extraordinary shift taking place across all of biology. New technology of all kinds was giving us more information than we ever dreamed of, allowing us to analyze it faster and faster, bringing us closer and closer to the fantastic beasts who share our world.

What did all this mean, and what might it add up to—this age of discovery, these machines of sensing and pattern recognition? If this was just the beginning, what else might computer algorithms find in all this new whale data?

At the end of the film shoot, I was visited by two young men from Silicon Valley. They'd made their fortunes founding internet companies, and now they wanted to help conservation. They told me something totally fantastical—that they wanted to take the most recent artificial intelligence tools and use them to decode animal communications, something "like a Google Translate for animals." I instantly thought back to what Joy had said when we were speculating as to Prime Suspect's motivation: "You can't just ask a whale, 'Why did you do that thing?' "

I thought of all I'd seen, and the radical change in what biologists could record and discover in their recordings. Why couldn't we just ask a whale? What was standing in our way? What, scientifically speaking, was so impossible about it? I decided to find out.

2

A Song in the Ocean

Only if you love something will you inconvenience
yourself to work on its behalf.
—Barbara Kingsolver

Over the years, as I immersed myself in the world of cetaceans, the people I met kept mentioning one man. For many, it was his research that had first touched them and drawn them to a life working with these animals. It was from him that I learned the power and importance of decoding cetacean communications. Dr. Roger Payne is the man who gave the whale its song.

It turned out that this iconic whale scientist lived far from the sea, deep in the woods of Vermont. On a Friday in June, I turned off the highway and down a long road that undulated through a forest. It was a beautiful day, the sunlight peppering the road through the trees. Bob Dylan croaked away on the car stereo, and my mind was wandering. A very large, dark dog appeared at the edge of the woods, up ahead to my right. I slowed down in case it should make a dash across in front of me, and it did. Only then did I realize that it was a black bear. It crossed to my side of the road and turned to look at the car for a moment, then swung its head around and ambled into the scrub alongside the car and onward down a slope, toward where an unseen stream ran. The rustling plant life marking its passage. A short distance later I reached a tall, timber-clad farmhouse painted white. There were beehives in a clearing to one side, and hummingbird feeders mounted against the windows. The house looked out across a grassy clearing and down over the rolling treescape, with no sign of other human habitation as far as I could see. I felt about as distant from the sea and whales as you could be.

I knocked at the door and a tall, beaming man appeared, dressed in a gray shirt and chinos and wearing wire-rimmed specs. He didn't seem eighty-three, and I blurted that out. "Oh no, I seem very much eighty-three if you're inside my head," he said, but his movements were quick and there was a youthful glint in his eye.

Roger insisted on showing me around before our interview. He led me around the back of the house and past the open doors of his workshop, where he likes to spend time building and repairing things made with wood. Roger explained that the house wasn't his but had been lent to him and Lisa, his second wife and an actress of some acclaim, by a whale-loving friend impressed by Roger's life's work studying and protecting cetaceans. Roger wasn't alone here: Hidden in the woods was a small lake complete with a floating teahouse, which was regularly visited by monks from a small monastery in a neighboring farmhouse. A large golden statue of the Buddha shimmered through a copse across the road. We walked down to the lake, where wagtails and other small birds hopped around its shore. A man was meditating on the floating wooden jetty of the teahouse. Next to the lake there was a long grassy ridge, almost ten feet tall. A tunnel penetrated the ridge, constructed from giant rough-hewn pieces of stone, beautifully worked together, its opening wide enough for two people to walk through. It was like the entrance to a burial chamber. All this was the project of his cetacean-loving benefactor. The tunnel was dark and cool, and opened out onto the other side of the ridge, where we emerged into a ring of colossal standing stones, a dozen or so rough obelisks each over ten feet tall. Roger told me he and Lisa had married here, in a ceremony that his friend, the author Cormac McCarthy, had helped him write, and which was read by another of their friends, *Star Trek*'s Sir Patrick Stewart. I felt I was in the presence of a benevolent wizard.

Back at the house, Lisa, a New Zealander homesick for Britain, fed us her freshly baked bread, bemoaning the impossibility of finding quality baked

Roger Payne listening to whales off South America.

goods or her favorite cheese in this part of rural Vermont. Roger settled into a high-backed, flower-print armchair, with a tabby cat in his lap, and started talking.

In the late 1950s and early 1960s, he was studying owls, specifically how the mechanics of their hearing allowed them to catch mice in total darkness. He was a gifted scientist and on track to continue this work. But then something happened that changed his life. While working near the Massachusetts coast at Tufts University, he heard an announcement on the radio one night. A whale had washed up on a nearby beach. Roger decided to drive down. By the time he arrived, it was dark and pouring rain and everybody else had left. Roger walked along the beach and found the body, noticing that it wasn't a whale but a dolphin. Somebody had cut the dolphin's tail flukes off, another had stuffed a cigar in its blowhole, and another had carved their initials in its side. Powerful waves beat the shore. Roger stood in the dark and the rain and looked at the dolphin, at its "lovely curves" softly illuminated by the glow from nearby buildings. He was deeply upset, and later wrote, "I removed the cigar and stood there for a long time with feelings I cannot describe. Everybody has some such experience that affects him for life, probably several. That night was one of mine."

In that moment, Roger realized that a person could only carve their name in a dolphin if they viewed it as something very different from a human, as something unknown, as no more than a *thing*. It was "nuts," he thought, "there's got to be a different way." But he felt powerless to change something so monumental, and resumed his studies. Sometime later, he attended a lecture given by a gentleman who ran the bureau of international whaling statistics. This man spelled out "the cold hard truths" about what was happening to whales worldwide: the unfettered industrial slaughter as the factory fleets worked their way down through the largest, most profitable, and most easily found whales—the right, blue, and fin whales—and when they'd hunted all of them, through the others—the sei, humpback, sperm, and minke whales. Roger was shocked.

A few days after Roger attended the lecture, he happened to hear a recording of the calls of right whales. Roger had never heard such a sound before, so mysterious and lovely. It haunted him. He programmed his alarm clock to turn on a record player that would play it each morning to wake him. "I figured, if I can wake up to these sounds, maybe the day will be better. And I did, and it was."

But the calls of the whales weren't just beautiful; they were a reminder of the dire straits they were in. To Roger, part of the problem was that the only connection people had with whales was through the whaling industry. Killing a whale as soon as we saw one was "not a terribly good way of inspiring people as to its complexity and interest and variability and clever wit or anything else." He became determined to change this. And one day, at a meeting of the New York Zoological Society (now the Wildlife Conservation Society), he simply announced that he was going to study whales. By his admission, he had no idea what he was talking about; in fact, he had never even seen a living whale. But encouraged by the positive reception of his announcement, he decided to press ahead with what would be a lifetime dedicated to these giant, persecuted, mysterious beasts. As with many whale scientists, studying whales was also an appealing way to live. "I've always wished to go down, you know, smaller and smaller roads until you get to a track that leads to the sea," he said. "And if you get into a boat at that point, and set forth, that is a kind of a living and a sensual experience, which I find irresistible." As he spoke, it was as if he put into words what I had also always felt but had never been able to articulate.

Roger began to cast his net. He spoke to anyone who knew anything about whales. Eventually, he got a tip that a U.S. Navy engineer named Frank Watlington in Bermuda had made some strange recordings. It was the height of the Cold War, and the United States had listening stations underwater to eavesdrop on passing Soviet submarines. Frank had access to a single hydrophone in a top secret array of listening devices on the seafloor, thirty-five miles offshore. This hydrophone had a broad range and picked up all the frequencies humans can hear. When he heard something interesting, he'd record it. He'd been hearing some unusual sounds—long, varied, and complex—and noticed they seemed to correlate with the times when he saw humpback whales migrating past Bermuda. He began to wonder if the whales could be making them. Roger and his then-wife, Katy, traveled to Bermuda and met Frank. The Navy engineer took them into the bowels of a boat where there was a generator hammering so loudly Roger couldn't hear himself speak. He put Frank's headphones on and listened to the recording. "I think it's a humpback!" Frank shouted at Roger, who, as he heard the sounds, was transformed. *If this is a humpback whale*, he thought, *this will speak to the world as no other voice has ever spoken to the world*. Interviewed decades later Katy remembered it vividly: "Tears flowed from our cheeks. You know, we were just completely transfixed and amazed." To this day, Frank Watlington's recordings are still some of the most beautiful and haunting humpback vocalizations ever captured.

It was 1967 and commercial whaling was at its peak, with more than seventy thousand whales killed each year. Watlington was concerned that whalers would use the sounds to find and kill more whales. Frank gave Roger a copy to take home and told the Paynes to "go save the whales." Roger listened to Frank's tape for three months, at all hours and every opportunity. The calls of the whales were highly complex, lasting about twenty minutes, ranging from harsh, belching grunts to high squeaks to deep, mournful moans. He listened to the record hundreds of times until it hit him: "My God, these animals are repeating themselves."

With a collaborator named Scott McVay, Roger made visual representations of the sounds called spectrographs, which clearly showed these repeating patterns. The patterns were made up of units ("notes") of different pitches and volumes. From rumbles on the lowest register of our hearing to screeches close to the highest notes we can perceive, these notes were produced in groups ("phrases"), which repeated for a few minutes, forming a "theme." The first recording on Frank's tapes consisted of six themes, Roger explained, and you could give each of these a letter (A, B, C, D, E, and F). They found each theme was repeated a variable number of times before the whale switched to the next. When the sequence cycled back to the first theme (A), Roger and Scott called the sequence a "song." The first song looked like this: *AAAAABBBBBBBBBBBBBCCDDEFFFFFFFF.* When the whale returned to cycle back to sing the A theme again, it marked the beginning of a second rendition. A humpback whale will not wait long for an encore, so "their bouts of singing are rivers of sound that flow on for many minutes and sometimes even for hours."

Most animal vocalizations are linear, meaning they do not have nested hierarchies in their structures. Roger, a cellist, felt that the closest thing to these whale calls was music—which is why he called them songs.

Payne and McVay's 1971 research paper was a blockbuster, with their spectrographs writ across the cover of the journal *Science*. Inside, they wrote: "Humpback whales (*Megaptera novaeangliae*) produce a series of beautiful and varied sounds for a period of seven to thirty minutes and then repeat the same series with considerable precision. We call such a performance 'singing' and each repeated series of sounds a 'song.'" Roger and his fellow researchers observed that just the bull humpbacks vocalized. They hang vertically, motionless, in the water some sixty-five feet below the surface and sing complete songs, one after another. At the end of a number of songs they return to the surface to inhale, and then sink down again to continue their calls. Usually, they don't interrupt the song to breathe until they reach a particular theme, but regardless of where they breathe they "quickly tuck their breaths in between the notes so as not to interrupt the performance of the song, just as humans do when we sing." And if they are uninterrupted, such song bouts can go on for many hours—even days. In creating their songs, they even employ some musical laws similar to what we humans use in creating ours. For example, humpback whales include percussive and tonal sounds in their music in about the same ratio that we do in several of our musical traditions.

A member of Roger's lab, Linda Guinee, along with Katy Payne (herself a highly accomplished musician and scientist, whose work on animal communication has stretched from whales to elephants), codiscovered that whales even employ rhyme. I asked Roger why this might be. He said that Linda, Katy, and he speculate that whales may use rhyme for the same reason the ancient Greek bards used rhyme in their epic poems—to help them remember what came next in a long song.

It was Katy Payne who led the way in showing that the songs sung by whales are constantly changing—something that is unusual for singing animals. Humpback whales live in a dozen or so known, distinct populations, each of which breeds and feeds in different parts of the many seas across the globe. They seem faithful to a given feeding area, but the males are known to visit several breeding areas. At the beginning of each breeding season, the whales of a population may all sing slightly different songs to one another. Over the season, like an orchestra tuning up, these seem to coalesce into a single, coherent song that is quite accurately repeated. These songs evolve continuously, each one changed from that of the previous year, until after a few years it would become entirely different. Various populations of whales sing distinct songs in different oceans, but there appear to be "hit factories" like the Australian whales, whose ear-worm tunes seem to leak out from their population, carried by bulls to other seas where other bulls take elements of their phrases and verses and add them to their own songs.

The songs appear to never repeat a previous theme pattern. Katy Payne quoted Edward Sapir, who studied human linguistics and drew a parallel in whale song with his description of how human languages change with time: "Language moves down time in a current of its own making. It has a drift... Every word, every grammatical element, every locution, every sound and accent is a slowly changing configuration." And it was not just humpbacks who sang. It turned out that other large whales, such as blue whales in the Indian Ocean, did, too, only their songs seem to be much simpler. Bowhead whales, which can live for more than two centuries, sing songs that have been compared to jazz. Researchers following on from Roger's work found that the seas were alive with the vocalizations of whales and dolphins, dizzying in their variety and ubiquity. Some could be heard only from a hundred yards away, while others could transmit across entire oceans.

The cat shifted in Roger's lap, and he settled deeper into his armchair. As he had been speaking, a shaft of light from the window had slowly crept across the bookcase behind him. This was the part of the story I was here for—what had he discovered? What did the songs mean, if anything? Why do whales sing them? Why are they so complex? Roger answered that no one knew. All of this was and is still the subject of fierce debate. Roger recognized that it was possible they might have no meaning at all—but he himself can't believe that. His suspicion, based on the huge effort the whales put into their performance and the fact they learn different versions, is that they must have some "incredibly important meaning." Yet he has all but given up hope of knowing what this meaning might be in his lifetime. Roger is betting on the possibility that the songs are in major part a mating display that males use to gain the attentions of females, as with birdsong. But he has several problems with this theory, not least that unlike with birdsong displays, the humpback females don't obviously seem to pay much attention to singing males. "Oh yes," he chuckled, "the more you study whales the less you know about their total, like physics." He paused. "I would desperately love to know what they mean."

I'll admit, this was a bit of a disappointment to me as a scientist—I yearned for some answers, and we had discovered that humpbacks could sing half a century ago! Roger told me the question had kept him awake at night for decades. His enigmatic discovery was just the start of his story. The songs he'd discovered were beautiful, but their singers at that time were in danger of being permanently silenced—363,661 whales were butchered in the 1970s. Roger had gone to investigate the songs not just to find out what they meant, but because he knew their power: to "capture the fancy of humanity."

Roger had been confident from the start that if other people heard the songs, they would think differently about whales. They would care.

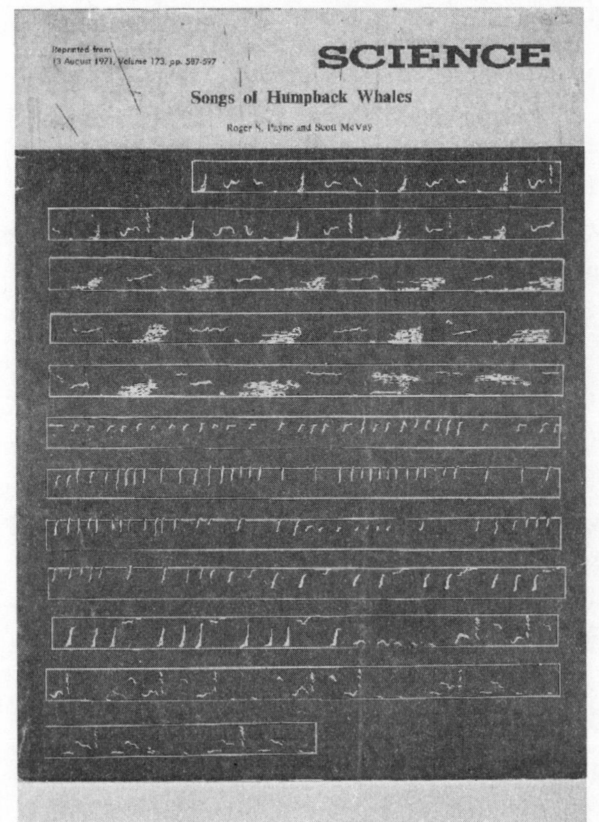

Payne and McVay's seminal paper, representing the "music" of whale song, in Science, *1971.*

In 1970, before he even published his scientific findings in *Science*, he released an album of the best recordings, *Songs of the Humpback Whale*. It sold 125,000 copies and went multiplatinum. He became a PR guru. He played the album to singers, musicians, churchgoers, actors, poets, politicians, and journalists, to anyone he thought might care and bring more attention to their beauty. He went on late-night TV and radio talk shows on both sides of the Atlantic. The "sounds spread like wildfire; people got into them, and when they heard them they were stunned." He heard rumors that Bob Dylan would sometimes stop gigs to play sections of the record. Whale song went viral. They were played on Johnny Carson's *Tonight Show*, on *The David Frost Show*, and in the background of Judy Collins's hit song "Farewell to Tarwathie." The cetacean song resonated with the growing environmental movement. Roger's album came out within months of the first Earth Day, and the year before Greenpeace was launched. Audiences who had already warmed to dolphins from watching the adventures of *Flipper* brought whales into their pantheon of beloved beasts.

The biggest coup came when Roger was able to persuade *National Geographic* to press a disc to include within the January 1979 issue of the magazine. At the time, the magazine's circulation was 10.5 million, so 10.5 million flexi disc pressings—with a selection of songs from humpback whales—were made. To this day, this remains the largest single print order of *any recording ever made*.

In interview after interview I undertook half a century later—with scientists, whale-watching boat captains, free divers, underwater camera operators, and other people who love whales and dolphins—I found it was this record that had spellbound them as children and teenagers and hooked them on whales for life.

As the numbers of surviving whales finally plummeted under the strain of centuries of relentless hunting, protests grew. Footage of whaling was aired on television wildlife documentaries. People whose grandparents had worn whalebone corsets donned T-shirts reading "SAVE THE WHALES." Greenpeace boats drove between harpooners and their quarries, playing Roger's album. Public pressure built up and turned into international political pressure. In 1972, the United States passed the Marine Mammal Protection Act, a law prohibiting the hunting and killing of whales in U.S. waters, and the import and export of their products. The International Whaling Commission (IWC) went from creating quotas for whalers to closing down all hunting. Finally, in 1982 a moratorium on commercial whaling was voted in. The hunting has now largely stopped. (As I write this, however, Japanese whalers, abandoning the pretense of hunting whales for science, have left the IWC and resumed their commercial hunt in their national waters, although few in Japan seem to want to eat whale meat.) Roger had wielded the songs with great power to save the whales—appealing not to our reason, but to our emotions and empathy. He gave the whale its voice in our culture, and this individual human decision is one of the reasons there are still whales at all.

As our interview came to its end and Roger stood up to get ready for dinner, dislodging the cat to some new warm corner, I realized something about his story. Roger had searched for something that could connect humans to whales. He heard the songs years before he published his landmark paper. To make that connection, to reach those millions of people, he had to *prove* they were songs. He had to find their patterns and show their structures. For Roger, it hadn't mattered as much in the end exactly what whale song meant; the fact they sang was enough to change their fate. He had to nail the science to stir the heart. Fifty years later, he continues his mission.

Growing up learning about the natural world through David Attenborough documentaries, I was spellbound by the story of Earth and the life on it. I wanted to do nothing more than witness and discover and explore it. But these were hard times for the carefree naturalist. I have lived only half an average human life span, but since I was born it is estimated that half of all the vertebrate animals have disappeared. In just a few thousand years, we've caused the loss of 83 percent of wild mammals and half of plants. We have replaced the diversity of life with the fewer species that can live in a humanized world. When I look across the rapeseed fields, car parks, and golf courses of my homeland, where temperate rainforests once stretched and great beasts roamed, I think of the Caledonian chieftain Calgacus, who said of the destruction wrought by his Roman enemies: "They make a desolation and call it peace."

Alive today are some 25 billion farmed chickens. Their biomass is more than double the weight of all the remaining wild birds on the planet added together; in fact, so many are killed each year that their bones accumulating in rubbish dumps are becoming a paleontological layer, a future marker of the Anthropocene. Of all the mammals left on the planet, by weight, 96 percent are human and domestic animals, such as cows, sheep, goats, dogs, and cats. As for the seas, we're told that by 2050 there will be more plastic in them than fish. This mega-death is unusual in life's history. As a wildlife filmmaker, like so many of my peers, I became a sort of nature war reporter. But I'd never really looked into whaling until my own run-in with the humpback in Monterey Bay. Before that experience, I'd naively imagined that most of the whale killing had happened in the nineteenth century, in Herman Melville's time, when industrial society ran on cetacean products, and cities were illuminated with burning whale oil and corsets were ribbed with baleen from their mouths. But once I started reading about whales, including new research drawing on combinations of whale DNA and the records of whalers, I discovered that most of the cetaceans ever killed were in fact slaughtered in the twentieth century, and many in my own lifetime.

Fossil fuel–powered steel boats known as factory ships were able to catch the faster, bigger whales, such as the blue and fin whales, which had outrun earlier whalers in their sailboats. These ships had the capacity to kill them at a distance with exploding harpoons before hauling them aboard, only to continue hunting while on-deck human teams and machines relentlessly rendered the slain titans. Later they would be manufactured into commodities such as dog food, fertilizer, lubricant, margarine, chewing gum, and typewriter ribbons. Recent evidence has shown this practice was still apace in my childhood in the 1980s, with Soviet fleets feeding Siberian fur farms on the flesh of giants from Antarctic seas. It's impossible to know the exact total, but estimates suggest in the twentieth century we killed around three million whales, more than 90 percent of many populations. This is thought to be the largest cull of any animal, in terms of biomass, in history.

Three million whales.

The blue whale, largest by weight and size of all animals in the history of life on Earth, had been hunted until just 0.1 percent of its population remained. The largest blue whale population, in Antarctica in the eighteenth century, was estimated to have numbered around three hundred thousand. When hunting ceased a few decades back (mainly because there were so few whales left that it was hard to find the survivors), there were thought to be around 350. It is hard to imagine this global level of slaughter, the equivalent to killing every human in the world apart from the residents of Bulgaria. It is dizzying to try to consider the scale of what there once was before we industrially harvested whales—not just the animals, but also their behaviors, their cultures, and their communications. "We do not know the true nature of the entity we are destroying," wrote Arthur C. Clarke in 1962. Back then, it seemed to those studying whales that they would go extinct and be as lost to us as mammoths and dinosaurs. They would become old stories for children and dreams, relics of the vanished world.

And yet they did not die out. Thanks to the efforts of Roger and his colleagues, and the millions of people who protested and forced nations to enshrine protections of whales into law, many populations of whales around the world today are rebounding and expanding. This is a counterpoint to the dangerous narrative of innate human destructiveness, which leads to apathy. It shows how we can change, and life can recover.

In Monterey Bay, where I had my run-in with the humpback, I was told by fishermen and whaleboat captains that even spotting a humpback was practically unheard of in the 1970s. Yet early anecdotes from the first Europeans to arrive at this coast tell of there being whales in abundance. Now, once again, enough whales visit to make the chance of one leaping onto a kayaker less remote, and their number and visibility in the bay sustains a whale-watching industry worth millions. The central Pacific humpback whale population is thought to be recovering to prewhaling levels. There are other encouraging examples. Reports in 2019 and 2020 from the Southern Atlantic seas of South Georgia, where whaling stations totally exterminated blue whales a century ago, show a sudden reappearance, likened by scientists to the whales "rediscovering" the islands and their waters.

Roger and Lisa invited me to stay for dinner, and we talked long into the night. He told me another story, the implications of which took longer to sink in. In 1971, the first of two Voyager Space Probes was launched. These spacecraft are science tools, equipped with cameras and sensors. But they are also messages, expressions of life on Earth sent into the void. Roger explained that each vessel carried a twelve-inch gold-plated copper disk onto which was encoded photographs, diagrams, and audio recordings of things that humans at the time thought were important. It is an eclectic mix: the sounds of waves breaking, photographs of humans eating food, engravings of our anatomies, a pictorial representation of how we reproduce.

There were greetings in fifty-five languages, from Akkadian to Welsh. The astronomer and public intellectual Carl Sagan and his wife, Ann Druyan, had been asked to compile the recordings, and they included

In this image, engraved in gold and sent far into space, are three humans engaged in demonstrations of licking, eating, and (spectacularly) drinking. Were the models told they'd be interstellar banqueting ambassadors?

Roger's whale songs. The whales come after an address from the secretary general of the United Nations and a selection of sounds and messages in different human languages. "I should like to extend the greetings of the government and the people of Canada to the extraterrestrial inhabitants of outer space," intones the Canadian delegate to the UN, and beneath his words fade up three minutes of the haunting and mysterious calls of the humpback whale.

Roger checked. Both probes were now over 19 billion miles from Earth, traveling more than thirty-four thousand miles per hour. They are among only a few human objects that we have ever sent beyond the gravitational control of our Sun and so have escaped our solar system. If, in five billion years, our species fails to get out of our solar system before the Sun comes to the end of its life and in its death throes engulfs Earth and other near planets, these probes may become the only record left of our human existence, and those recordings may become all that remains of the whales. But Roger does not think any aliens that receive it would be impressed. If it had had, say, "sixty-two greetings from animals and one from a human," then the aliens would be justified in thinking us advanced. But our focus on ourselves, our sidelining of our fellow passengers on our planet, is for him evidence that we have only just "put our toe on the lowest rung of the ladder... to a point where we can proudly announce to the court of intergalactic opinion that yes, there is intelligent life on Earth." And at this he guffawed.

Then he continued, in deepest seriousness, "But when you think, who is that message on those two golden records really for? I think it's for us. That's my feeling." If the whales could give us an understanding of the world, of an empathy toward other species, then it would be the single most important lesson we could learn. As I readied myself to leave, he told me that the dominant problem we face is what we are missing by not being in closer

Humpback whale, Revillagigedo Archipelago National Park, Mexico.

touch with the rest of life on Earth. "If we are to have any future at all," he said, "we have to somehow make sure that what we do is preserve the rest of life on Earth, because without it, we won't make it." For Roger, showing people the world over how to empathize with whales was a critical bridge to a cultural change essential to our own survival—a way for us to comprehend our interconnectedness and correct our wanton course.

I'd embarked on a journey to try to understand *how* we might speak whale and decode animal communications—but after spending the day with Roger Payne, it dawned on me that *why* we might want to do this was of much greater importance.

3

The Law of the Tongue

Imagine the possibilities... the things we might see through other eyes, the wisdom that surrounds us.
—*Robin Wall Kimmerer,* Braiding Sweetgrass

Over the months observing humpback whales in Monterey Bay, I noticed something peculiar. Sometimes, when the whales perceived our boat, they seemed to want to avoid us, and would simply disappear. At other times, they appeared to take no notice at all, and ignored us. But on some occasions, the whales would approach.

As per the rules, we would stop a good distance away, and point our boat in the opposite direction from the way the whales were swimming, and yet they would change course and come straight toward us. They would poke their heads out of the water and stare at us with one eye, before rolling over to look at us with the other, then swim around the boat, seemingly examining it, splaying out their giant pectoral fins and twisting their bodies. They would carefully rise to the surface right next to us, within touching distance, and exhale and bob there unmoving, a behavior known as "logging." At times, the whales would seem to display by gently splashing their fins, holding their flukes in the air and wobbling them back and forth, while their heads were pointed down, something I like to do with my feet when walking on my hands on the floor of a swimming pool to wave at my friends' children. It's unclear what humpbacks would gain from these encounters. They are not fed or helped by the boats, or sheltered from foes. It seems unlikely the displays are threats to the vessels. They didn't match how humpbacks show aggression—exhaling loud bellows, or slashing their pectoral fins and tails at threats, as I saw them doing toward killer whales—and the encounters never escalated to violence. Whale watchers call these individuals "friendly whales"; they stand out from their peers so much that when the captains recognize one of the "friendlies," they'll make sure to wait around for one of these enigmatic interspecies interactions.

In the wondrous variety of life on Earth, there are many relationships where different species interact. These are called *symbioses*, which translates as "living together." These various interactions are separated by biologists into categories based on who gets the best outcome. In 1975, a biologist was diving in Indonesia and scooped up a leopard sea cucumber. Sea cucumbers are leathery marine creatures resembling an animal made from one arm of a portly starfish. He put it in a bucket of salt water. A moment later, to his astonishment, over a dozen fish swam out of its anus. Star pearlfish are svelte, slippery, mostly defenseless fish that like to hide inside other animals. Sea cucumbers breathe through their anuses. This has many advantages, but one major drawback. If you need to breathe via your anus you cannot keep your sphincter clenched shut forever, even if a star pearlfish is around. The diver counted fifteen star pearlfish as they emerged from this single leopard sea cucumber's rear end, each of them at least a quarter of the length of the poor invertebrate.

A star pearlfish admiring the view from a sea cucumber's anus.

Scientists do not think the hapless sea cucumber benefits from having fish living up its anus, and some species of pearlfish are terrible guests and devour the reproductive organs of their host. To defend against this, some sea cucumbers have evolved "anal teeth." This puts these sea cucumber–star pearlfish relationships into the *parasitic symbioses* category, where one individual benefits to the other's cost.

Then there are *commensal symbioses*, in which one individual benefits with no clear advantage or harm to the other, such as when birds called cattle egrets hang out with cows to eat the insects disturbed by their grazing. A kind of win-"meh" dynamic. Some species of barnacles (sticky crustaceans related to crabs) have evolved to settle onto the skin of whales, where they cement themselves, grow a hard protective shell, and enjoy their free ride, filtering food from the passing water. Some whales have been found carrying almost half a ton of barnacles.

Finally, there are *mutualistic symbioses*—and these are my favorite interactions. Mutualistic symbioses are the Disney-esque, win-win partnerships where different species team up for mutual benefit. For example, in South Africa, there is a bird called the honeyguide. It is great at finding beehives and loves eating the wax and larvae, but its size prevents it from tearing into the hive and defending itself against a swarm. To this end it enlists the honey badger, a badass, four-legged, black-and-white mustelid with thick fur and tough claws who loves eating honey. The bird guides the badger to the hive, the badger tears it open, shrugs off the bees, and eats the honey; the bees disperse, and the honeyguide swoops down to peck at the delicious honeycomb strewn on the ground. When excavating a beehive, the honey badger turns its anal pouch—a powerful, foul-smelling sac—inside out, which is thought to fumigate the swarm. A biologist named Pocock writing in 1908 described the smell as "suffocating," which it appeared to actually be for the bees, "which either flee or become moribund" when the honey badger lets rip. As a former beekeeper myself, I envy the honey badger its pouch.

Mutualistic symbioses can sprout up between distantly related animals. On the seabed, invertebrate pistol shrimp team up with vertebrate goby fish. The seabed is a dangerous place where vigilance is required to spot predators, and a nice, deep burrow can provide vital shelter. The goby is not good at digging, but his eyesight is far superior to his shrimpy associate and will alert her to approaching predators and other dangers she might otherwise miss. The shrimp will dig into the sand and excavate a system of large burrows, much bigger than a solo shrimp would need. There are side tunnels for her to shelter in when she needs to grow and shed her shrimpy exoskeletons, but also wide ones large enough that her roommate goby can seduce and mate with another fish in them. (Clearly, this is a close association.) Fish and shrimp form their alliances when juveniles, growing up together. If the goby fish is in the burrow and it collapses on him, he will not panic but will calmly wait for his friend the shrimp to dig him out and repair their hideout.

Other examples of mutualisms are everywhere and aren't restricted to the animal kingdom. You may have seen lichens sprout on old walls or gravestones. These appear to be one living organism but are actually formed of two or more very different creatures from different domains of life. The lichen is a mixture composed of both a fungus (kingdom Fungi) and either algae (kingdom Protista/Plantae) and/or cyanobacteria (kingdom Monera). The fungus creates a structure and home for the algae or cyanobacteria, and in exchange they turn sunlight into food, which the fungus eats. Together they form a completely codependent composite organism made up of two or more species so distantly related they last shared a common ancestor billions of years ago (you shared a common ancestor with a whale just 145 million years ago, for comparison).

Yet another example of cooperation between kingdoms of life is found on acacia trees. These trees sometimes develop galls on their bark: woody chambers that are ideal homes for certain ants. The ants colonize the galls, and when a browsing giraffe approaches the tree to gorge on its tender leaves, the tenant invertebrates rush to the scene to defend their landlord, squirting acid at the giraffe until it is discouraged.

Interspecies communication is an integral feature of life on Earth and has been around for billions of years. All these mutualistic symbioses have one thing in common: They are held together by signals. A growing fungus will send out special feelers called hyphae and produce mucus to sense the signaling molecules on potential algae teammates, in order to size them up with a view to making a lichen together. The honeyguide bird sings a special song to the honey badger to get its attention, then flies ahead to lead it to the beehive. A foraging shrimp will keep one of its long antennae resting on its goby pal's tail so that if the eagle-eyed fish spots danger, it will signal to its myopic friend by waggling its tail and both will scuttle to safety. An acacia tree will release chemical signals (hormones) that alert its resident ants to a munching herbivore and tell them where to come to help. Living things survive by signaling to other life-forms, within and across the species boundaries. This includes both whales and humans.

Almost everywhere we go, humans have developed associations with the other animals around us. We have learned to interpret their signals, and they ours; to seek shelter together; to find food; to protect each other. We steer their movements and they indicate important things we should pay attention to. Getting these signals right has often been a matter of life or death for both human and beast. Sometimes we signal explicitly, as when a shepherd sends his dog across distant fields and guides its steps, telling it with ritualized whistles to run wide or crouch down, to herd his sheep back to a tiny pen. Other times, the signals are subconscious. Recent studies have found that horses can sense the heart rates of their riders through their skin, and that they respond to their stress; a horse's heart rate and stress level increases or falls in tandem with those of the human on its back. For thousands of years, these relationships have been mutualistically symbiotic in nature: There has been something to gain for both parties in understanding the signals of the other.

In many cultures, there are those whose job it is to pay attention to nature for signs and omens. Milkmaids, shepherds, wolf hunters, pigeon racers, rat catchers, otter fishermen—all have paid very close attention to their animal associates. There are many anecdotal stories of humans and other animals in mutualistic symbiosis, and signals are at the heart of them all. Some can be explained by training, where we have rewarded or punished an animal for exhibiting a certain behavior, known as "operant conditioning." In these circumstances, there is no need for the animal to understand why it gets the reward, only to respond as it did previously when it was rewarded. In Piauí state, Brazil, a parrot who had been taught by its crack dealer owners to shout "Mum, the police!" was taken into custody when the cops arrived. According to the *Guardian*: " 'He must have been trained for this,' one officer involved in the operation said of the two-winged wrongdoer. 'As soon as the police got close he started shouting.' " Since his arrest he has kept quiet.

Other stories are a little harder to dismiss as simply operant conditioning. One famous historical human-animal team was James Edwin "Jumper" Wide and his friend Jack the chacma baboon. Jumper, a nickname derived from his habit of leaping from one railway car to another, was a railway guard who lived in Uitenhage, South Africa, in the 1880s. But then he lost both his legs below the knee after falling beneath a train. Not long after this accident, he was rehired as a signalman at the Port Elizabeth mainline railway outside Cape Town. According to the story, Jumper was at a market, where he saw a young baboon that had been trained to lead an ox wagon. Seeing his potential, he bought the primate off his owner and named him Jack. Soon Jumper had trained Jack to be his apprentice signalman and to pull him around on a wheeled sled. At the station, a number of levers pulled sections of track backward and forward to send trains along different routes. The two lived together in the signalman's accommodation. This suited both Jumper and Jack very well.

To teach Jack the appropriate lever to pull, Jumper invented a signaling system: He would point to the number of the lever by holding up the corresponding number of fingers. This system worked seamlessly, with Jumper giving Jack a few sips of brandy each evening to keep him sweet. (Approaching trains would whistle to indicate which levers to pull in which order.) So far, this could be explained away by operant conditioning. But what was more interesting were the reports that Jack was soon able to interpret the trains' toots for himself and, upon doing so, would leap into action, pulling the right levers in the right order to send the approaching locomotives along the correct tracks.

The baboon also learned to respond to other signals. If an arriving train gave four blasts of its whistle, it indicated the driver needed a set of keys from a special box. Jack observed Jumper hobbling on his prosthetic legs to the key box at this signal and learned to spring up ahead of his friend and fetch the keys for him.

Jack and Jumper at their signal box, circa 1885, with Jack pulling one of the track levers. The trolley that Jack would pull Jumper around on is parked to the right.

Trouble loomed one day when a passenger was distressed to see a monkey operating the railway signals and, following an investigation by the railway company, the team of man and baboon was fired. Fortunately, after other employees appealed, the company decided to test Jack. They confronted him with a series of improbable and fast-changing train whistles in a mocked-up signal station like his own. He was found to be so good at his job that he was not only allowed to continue but was also given monthly rations and an employment number from the government. "Jack the Signalman" operated the signals for nine years without making an error and became something of a tourist attraction until he died of tuberculosis. Could all this have just been the mechanical mind of Jack responding to noises, pulling levers, and learning through reward or punishment what action sequence to perform next time? Or could he have somehow worked out the cause and effect, wanted to please Jumper, and truly understood what he'd been taught?

This is not the only story of baboon-human partnership. The Namaqua people of Namibia have long trained baboons as goatherds. The monkeys would follow their goats by day, guarding them. Then in the evening they'd bring them together and shout in alarm if a predator was spotted, and herd the goats home to their compounds at dusk, the goatherd sometimes riding on the back of the largest goat. This persisted at least to the 1980s. Some, such as one named Ahla, would also groom the goats, and she knew which kid belonged to which nanny goat, reuniting them when separated.

Perhaps the baboons had been unconsciously conditioned or consciously trained. Perhaps Jack needn't have understood anything except that certain levers must be pulled in response to certain whistles and trains, and that food and shelter would come as a reward. Ahla might have been responding to her hierarchy-trained instinct to return an infant to its mother. After all, these were semidomestic animals, raised from birth around humans and our strange ways.

After my encounters with the "friendly" whales, I went on the hunt for examples of human-cetacean mutualistic symbioses: times where people and wild whales had teamed up. And I came across one mutualistic symbiosis that defied easy explanation, where it's unclear who began the interactions, and who trained who. This, then, is the tale of the Killers of Eden.

Whales, dolphins, and porpoises all belong to one bunch of closely related animals (an infraorder) called the cetaceans, a name derived from *kētŏs*, the Ancient Greek word for "huge fish" or "sea monster." They are not fish but mammals; like us they have warm blood, breathe air using lungs, and give birth to live young whom they nurse with milk. Sometime around 50 million years ago, perhaps near modern-day Pakistan, some mammals began to move back into the water. These were the ancestors of all cetaceans. They lost most of their hair and whiskers and became streamlined and insulated with blubber. Their hands and feet turned into paddles. They gradually became so perfectly adapted for living in water that they could not survive out of it. They spread across Earth's seas from the tropics to the poles, to the bottom of the deepest oceans and up rivers far inland. Today there are at least ninety species of cetaceans. All are carnivores, consuming other animals for the nutrients and water they need to survive. In this book, I often use "whale" as a shorthand for all of them—whales, dolphins, and porpoises.

Cetaceans are classified into two kinds according to what is in their

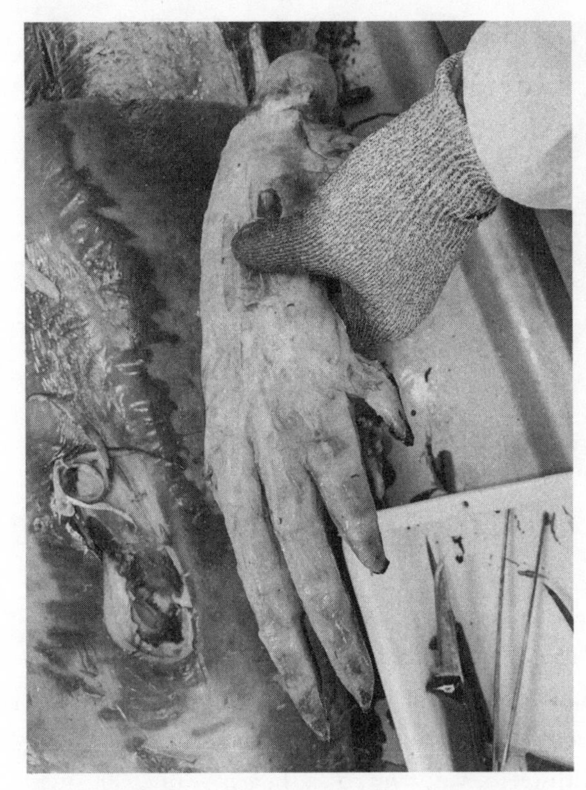

Inside every whale and dolphin flipper is a limb that first evolved to walk on land. Here the hand of a Sowerby's beaked whale, Mesoplodon bidens, *is held by the hominid hand that dissected it.*

mouths. There are the toothed whales, (or *Odontoceti*, literally, "sea monsters with teeth"); and there are the baleen whales (or *Mysticeti*, "sea monsters with moustaches"). These split from the toothed whales some 34 million years ago and have replaced their teeth with giant flexible bristly combs made out of a substance called keratin—the same material as your hair and fingernails. The baleen whales tend to take big gulps of seawater, from which they sieve their prey of fish and krill. They're generally pretty big. The humpback whale that almost killed us was a baleen whale, as are fifteen other species including blue whales, gray whales, right whales, fin whales, and minke whales.

Toothed whales, as their name suggests, have teeth. They can't feed by filtering massive gulps of seawater, so they hunt animals they can bite. All dolphins and porpoises are toothed. They range in scale from the dog-sized and critically endangered (there are perhaps just ten left) vaquita porpoise,

A humpback whale using its baleen to sieve fish while lunge feeding.

hunting tiny fish in the Sea of Cortez, to the apartment-sized sperm whale, which stalks similarly titanic prey—thirty-foot-long giant squid. A simple way of thinking about the two different hunting strategies that teeth and baleen give whales is that if you filter-feed, you eat small animals in big gulps, and if you hunt with teeth, you tend to eat bigger animals in small bites. Perhaps the most famous toothed whale is the killer whale, *Orcinus orca*. Some kinds, or "ecotypes," of killer whales hunt fish like salmon or herring. Other ecotypes hunt marine mammals, including some that specialize in hunting whales, even enormous species like blue whales. One theory is that their name derives from the Spanish whalers' term "ballena asesina," or "assassin whale." In some parts of the world, people today use the term "orca" instead, feeling "killer" is pejorative.

Baleen whales try to avoid the ecotypes of killer whales that hunt them. But their annual migration routes draw the baleen whales and their young calves through underwater badlands, where killers can lie in wait.

One of these places is the East Coast of Australia, along which southern right whales and humpbacks swim on their journeys to and from their feeding waters. There is evidence for human occupation across much of Australia dating well over forty thousand years, and some modern Aboriginal societies are thought to be the longest continuous cultures on Earth.

People here have stayed in the same places, and though they lacked written languages, their oral traditions have proven extraordinarily resilient. In some communities, there are names and stories about coastlines and landscapes that disappeared underwater after the last ice age. Descriptions of places retold in stories today match scientific reconstructions of the landscapes as they were ten thousand years ago, evidence that these tales have been passed down accurately for some four hundred generations.

Among the Aboriginal people of this coast, the Yuin nation, there were many beliefs, practices, and ceremonies linking the people with the whales. The black-and-white-patterned ceremonial dress of warriors resembles the body markings of the killer whales. One traditional cure was to crawl into the body of a dead whale and, but for the head, lie within its decomposing carcass. On the hillsides where people went to learn, instructive rock engravings still show the whales, one with the figure of a man inside.

In what the Europeans named Twofold Bay, by the colonists' town of Eden, the Katungal (saltwater people) had, perhaps for thousands of years, developed and maintained an extraordinary mutualistic relationship with the killer whales. Between April and November, killer whales would lie in wait for migrating baleen whales (or "Jaanda"), whom they would trap in the bay and devour in the shallow water. Here the Katungal could also more easily spear the baleen whales and use the meat. One theory is that this was interpreted by the people as the whales bringing them gifts. The killer whales became known as "beowa" (brothers), and the people are said to have regarded the killers "as the reincarnated spirits of their own departed ancestors." According to oral tradition and noted by early Europeans, the Katungal would reward the killers with the mouthparts of the prey whales, including their massive tongues, which could weigh up to four tons.

One hundred fifty years ago, a whaling settlement was established in the bay. The settlers worked out of small, shore-based whaling boats, ready to feed their societies' hunger for whale oil. Many of the European whalers considered the local killer whales their competitors, a nuisance. But one family of Scottish whalers, the Davidsons, hired Yuin to work on their boat and paid them a fair wage. They, in turn, taught the Davidsons how to hunt *with* the killers. The whalers got to know fifteen to twenty whales by their "saddle markings" (just as modern whale scientists do) and gave them names like Stranger, Skinner, and Jimmy. Many of these would have been female whales. Killer whale societies are not dominated by the males, despite their greater size. Instead, they are matriarchal, led by one or more dominant females and their matrilines: daughters, sons, and grandchildren. Female killer whales, like humans and elephants, experience menopause—it's thought this allows them to focus on leadership, using their lifetimes of experience to guide their pods. The present-day southern resident orcas off the North American Pacific Coast, for instance, are led by a female, L25, believed to be at least ninety-three years old. The pod that hunted off Eden was likely no different. Local Aboriginal whaling families and the Davidsons knew many of the killer whales individually by sight and personality. One whale in the pod that the whalers interacted with a great deal in the early twentieth century was a huge bull, Old Tom, who was easily distinguished from the others by his massive dorsal fin and "playful nature." Perhaps he was taught to interact with the whalers by his grandmother.

The story goes that when the pod Old Tom belonged to encountered humpback and right whales passing by, they would herd them into Twofold Bay, where the Davidsons lived. Old Tom and other whales would break away from the hunt and alert the humans by swimming right to the river mouth by their houses and breaching—throwing and slapping their tails on the surface—at all times, even at night. The Davidsons and their crews would rush out to their boats and paddle out to the killers, who would then guide the men to the prey whale, helping to corral and attack them until the whalers had harpooned and killed their prey. Sometimes the killers would even help by pulling on the ropes leading from the harpoons, tugging the ensnared whales toward the whalers' boats. According to Percy Mumbulla, the nephew of one of the whalers, "the killers would let them know if there were whales about" but the communication was two-way: "Ole Uncle would speak to them killers in the language."

Paintings, diaries, photographs, and etchings depict these multispecies maritime battles, with the fifteen-foot boats of the whalers dwarfed by their colossal quarry and the giant killer whales weaving and leaping around them. When men were knocked in the water during the hunt or when their ships were sunk, the whales would swim around them to protect them from sharks.

When a hunt was complete, the Davidsons' teams would attach the dead whale to a buoy and the killers would take their share of it, eating its huge, fleshy lips and tongue. It's thought the Davidsons were taught this by their

In this photograph, taken at the turn of the century, the whalers are looking ahead to the right of the frame, past their harpooner and out of the shot, to a humpback mother they are hunting whose calf trails her next to the boat. In the foreground is the massive dorsal fin of a killer whale.

indigenous crew. The whalers would then take the rest to be rendered for its valuable blubber to be used in soaps, fuels, and the manufacture of leather. It was also a good deal for the killer whales, who would normally have to spend many dangerous hours hunting a baleen whale by battering it with their tails, pushing it beneath the water or biting at vulnerable parts. This exchange, a formalization of what was perhaps a mutualistic symbiosis millenia in the making, became known locally as the "Law of the Tongue."

Using contemporaneous pictures of the hunts and diaries, it's been estimated that killer whales participated in the Eden whaling industry for more than seventy years, from the 1840s to at least 1910, alongside three generations of the Davidsons. When one of them, Jack Davidson, drowned with two of his children, the men were said to have searched for his body in vain for a week. Old Tom remained for this time in one small corner of the Bay, which was where Jack's friends found his body.

Cooperative hunting and many other interactions between human and whale were frequently recorded and even filmed. "I don't think there's been such a combination of trust and friendship between, certainly not anything in the sea and humans," said eyewitness Alice Otten (age 103 years), interviewed in 2004. But in the early twentieth century the whales disappeared. It is thought that Old Tom's pod was slaughtered by incoming Norwegian whalers in a nearby bay, unaware that they were firing on allies. At the same time, many indigenous Australians were being dispersed—moved from their traditional lands, taken away to schools, their old ways forbidden.

Finally, the only whale left in the bay was Old Tom, who reappeared in 1923. In a poignant coincidence, it was George Davidson whom he encountered. George was out fishing with a friend, Logan, and both were surprised to see Old Tom—even more so when Tom drove a small whale toward George's modest boat. George had his harpoons with him and speared the whale. Whales were few and far between by this time, and with a storm closing in and fearing that it would be the only kill of the season, George's friend Logan tried to pull the dead whale away before Old Tom had had his "share," with George objecting fiercely. A tug of war developed between the whale and the man, during which two of Old Tom's teeth broke off. Old Tom, with no surviving members of his pod, stood little chance. Logan's young daughter, who was with them that day, remembered her father aghast, saying, "Oh God, what have I done?" An ancient contract had been violated.

How did this mutualistic symbiosis begin? How was it developed and signaled? Whales and dolphins have fingers, but they are hidden deep within their rigid pectoral flippers. Their faces are fixed, lacking the muscles humans and baboons use to pull our features into helpful visual signals of different emotions and intentions. We live separated not just by our biology but by our medium—cetaceans in the sea, and humans on land. Yet despite all these hurdles, whales and humans learned to communicate, team up, and violently bridge their worlds.

As time passed, few outside Australia knew of or believed the stories of the Killers of Eden. The thought that orcas could signal to and cooperate with people became absurd. In fact, until the 1970s, killer whales were

In this photograph, taken in 1930, you can see George Davidson, sitting on the body of Old Tom, in Twofold Bay. The last of a dynasty of whalers, on the last of the whales they hunted with.

widely considered dangerous beasts. U.S. Navy manuals warned their divers that a killer whale would eat a human on sight. Until the 1960s, it is said Coast Guard helicopters practiced their machine-gunning on pods of wild orcas. Through the seventies and eighties, wild orca populations were decimated in places like the Pacific Northwest, with calves taken from their pods to be housed in amusement parks, to the horror of the First Nations communities whose lives and beliefs were also intertwined with the whales. Many, many whales were killed in this process, which continues in some countries today.

My research revealed other stories of human and cetacean symbiosis taking place, and some much more recently. In Brazil, bottlenose dolphins off Laguna chase mullet to the shore, where fishermen await them in the shallows. The fishermen cannot see the fish underwater, so they rely on the dolphins to signal them. When the dolphins slap their tails, the fishermen throw their hand nets. The dolphins seem to benefit from this by catching disoriented fish, and the fishermen catch more, and larger, fish than they otherwise would. An intriguing study found that the whistles of the individual dolphins that hunted with humans sounded different from their compatriots who didn't. The cooperative dolphins' vocalizations stayed consistently different whether they were with humans or just other dolphins, so it's not believed that the whistles are directed at the humans. One of the authors of the study suggested it was a way "for dolphins to label themselves as members of a particular social group." Reading this, I was reminded that not all humans seek out cetaceans, but it's often easy to spot those who do: They have dolphin tattoos and humpback earrings, they don orca T-shirts and beluga baseball caps, signaling to other humans their membership of the cetacean-loving clan.

One day, the online algorithms that noticed I like stories about cetaceans brought to my attention an update about a pod of wild humpback dolphins in Queensland, Australia. The dolphins are normally fed there by people who queue up by a cafe, habitually interacting with them a great deal. During the local lockdown precipitated by the COVID-19 pandemic, the dolphins had been starved of fish and human contact for weeks. The animals had been arriving at the shore bearing "gifts" of sea sponges, bottles covered in barnacles, and pieces of coral. What ideas about the world and humans, about cause and effect, about other minds and what might compel them to give you fish, fizzed in those dolphin brains? What, exactly, drove this behavior? Whose idea was it? Where did they learn it? Were they just hungry? Or lonely, too?

The more I looked at the scientific literature and the news, the more I was struck by how keen on interspecies interaction cetaceans seem to be. Pilot whales are attracted to the calls of the fish-eating kind of killer whales (who are not dangerous to pilot whales) and will swim toward them to hang out together. False killer whales in New Zealand seem to form "friendships" with common bottlenose dolphins. It turns out these are not random or fleeting events, nor opportunistic teaming up. Scientists found that individual dolphins partnered with individual false killers stuck together for over five years, traveling together for hundreds of miles. The animals, so different in size and shape and diet, would swim side by side on long ocean voyages, their lives intertwined. In Ireland, there's a solitary dolphin that regularly approaches boats and has made friends with one of their captain's dogs. A lost mother and calf pygmy sperm whale appeared doomed in 2008 when trapped behind a sandbar at Mahia Beach in New Zealand, stranding themselves repeatedly even when humans refloated them. Then a local bottlenose dolphin called Moko seemed to intervene, swimming up between the humans and the whales. The whales immediately followed Moko through a gap in the sandbar and safely out to sea.

Humpback whales have recently been found to come to the rescue of other species when they are being hunted, mostly by killer whales. More than a hundred incidents have been recorded where humpbacks charge in to protect not just their fellow humpbacks but other species from being attacked—defending other whales, dolphins, seals, and even giant ocean sunfish from their predators; the humpbacks put themselves between predator and quarry, lifting seals and sea lions out of the water on their backs, away from their hunters. In Monterey Bay, I witnessed a pair of humpbacks fight off two pods of killer whales trying to feed on a gray whale calf they'd killed. The humpbacks spent days protecting the body. It is not clear what the humpbacks gained from all these exhausting interactions, which carry considerable danger for them. Are there warring sides in the sea?

In some ways, then, the idea of collaborating with another species is not madness, for it already happens every day. The world is held together by mutualistic symbioses. Cooperation has been argued to be as important a force in evolution as competition. But teaming up for mutual gain, slapping the sea, and sharing food is one thing. How about something that we humans perhaps treasure more—a deeper connection, actually understanding the mind of another? While researching the story about the Killers of Eden, I came across a recording of an interview with "Guboo" Ted Thomas, made near the end of his long life. Guboo was a child of one of the Aboriginal whalers in the earlier story, born at the turn of the twentieth century. He spoke of how he witnessed his father and grandfather being "summoned" by the killer whales to sea to hunt, sometimes from their sleep. But what fascinated me most was a different tale with a different cetacean, of how his people would "sing dolphins in" to ask for their help. As a child, he'd gone to the shore with his grandfather, who noticed a large shoal of fish. Guboo's grandfather ran down to the water and clapped sticks together and danced and sang. After a long time, dolphins appeared and drove the fish to the shore and out of the water, where the men caught them—a reverse of the relationship with the killer whales, with the humans as the signalers. One detail in the recording lodged in my head. Guboo described how after the hunt was done, his grandfather had walked out into the ocean and stood with the water up to his waist. A big dolphin swam to him and put his head on top of Guboo's grandfather's arm. The man patted the dolphin and spoke to him, and "then the dolphin went chi-chi-chi-chi-chi-chii-chi-chi-chi-chiiichiii. He was talking to Grandfather, and Grandfather talked to him." The dolphin swam off, somersaulted twice, and was gone.

I would have loved to have seen this, to have recorded it. But it is a story—and like many of the other anecdotal stories in this chapter, as scientific evidence it is weak. Could he have communicated with the dolphin? Could anyone ever really "speak" with a cetacean? I needed to move from my story and others to the more concrete world of data, of facts, of things you can see and touch and measure. What can we deduce about the communications of cetaceans from their bodies, brains, and behaviors? In the words of Matt Damon in *The Martian*, it was time to science the shit out of this.

4

The Joy of Whales

The leviathan…
Will he speak soft *words* unto thee?
—*Job 41:1, 3* KJV

When Prime Suspect erupted out of the sea in Monterey Bay, the main thought I remember having was how fleshy it was. I could see the grooves and blemishes of its skin and the barnacles stuck to it. At a distance, whales look slick and smooth, almost abstract, but when they are very close, they resolve into great breathing, stinking animals. As it hung above me, it was clear that this absurdly massive and otherworldly object was a living, thinking, feeling being. A giant animal filled with blood and bone, riven with nerves, up in the air above us.

I knew this intimately, because as well as being landed on by a falling whale, I have also had the fortune to meet a dead one. I had seen inside it, traced my hands along the joints of its bones and felt its warm heart. I owe this second honor to Professor Joy Reidenberg, the scientist who believed, after watching the video clip, that Prime Suspect had altered its breach in an attempt to avoid hitting us. It was her "You can't just ask a whale" comment that had spurred me on this journey. Joy is one of the most extraordinary people I've met, who also happens to have one of the world's most disgusting jobs.

In 1984, Joy was a fresh-faced grad student who, while speeding up a highway, was stopped by a state trooper. Joy didn't notice that he was on edge when he requested her ID and documentation, but she did know things would get weird if he looked into the back of her vehicle. When the inevitable moment came to pass, Joy waited in horrified silence. The officer stepped away from the vehicle and, with one hand on his service weapon, asked her to explain what he had found inside. "It was just my stuff. Bone saws, skull chisels, hammers, daggers, a sickle, garden shears, flensing knives, gaff hooks, trash bags, both chainmail and thick rubber gloves, and overalls," Joy said with a chuckle. For the officer, it must have been terrifying: A human body had recently been found chopped into pieces and bagged up. A killer with the tools and knowledge to disarticulate a person was on the loose, and he thought he had just found her.

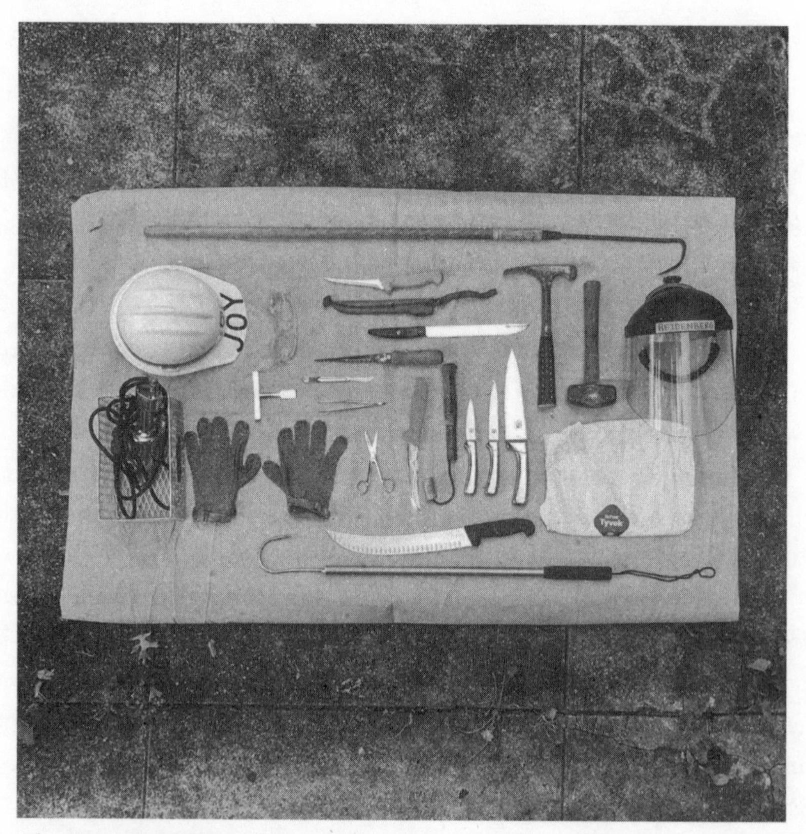

These are some of Joy's dissection tools.

She explained to the officer that she was on her way to her first assignment. A pygmy sperm whale had washed up three hours' drive away and she had been tasked with collecting specimens from the carcass and to perform a necropsy—an animal autopsy—to determine why it had died, measure its body, and take tissue samples. To save what could be saved and studied. Fortunately for Joy, her story checked out. The cop, impressed and no doubt relieved, drove ahead of her with his siren blaring, clearing the traffic and escorting the marine detective to her crime scene.

Joy had good reason to rush. A dead cetacean will decompose very fast. Unlike seals, whales have mostly lost the hair their ancestors needed on land. Some species, such as humpbacks, have retained whiskers along their jaws and snouts, useful for sensing the world around them. This may seem weird, but we have lost much of our body hair, too; in both whales and humans, our developing fetuses often have a hairy stage, a hint of our common hirsute past. In place of a cozy pelt, whales are kept warm and insulated by a thick layer of fat called blubber, directly beneath their skin and around their entire body, like a sleeping bag made of butter. Once an animal has died, the process of cell death releases heat. In the case of cetaceans, this heat is trapped inside their blubber, and they quickly cook themselves. Depending on the air temperature and exposure of the body, their brains, organs, and other soft tissues can turn to goop within a few hours, and all the information a racing anatomist is looking for is lost.

Joy's fascination with the inner workings of marine mammals and what they can explain about their outward abilities and behaviors has given her an almost unparalleled understanding of cetacean hardware—including their anatomy of communication. Perhaps the first thing to think about when it comes to the idea of decoding cetacean communications is this hardware: What clues does a whale's body hold about how they might think, listen, and speak? Nobody was better equipped than Joy to help me find answers. She has lost count of how many whales and dolphins she has dissected (hundreds, she thinks), and it was she who had first shown me around inside a whale. It was on a cold beach on the southeast coast of England, in March 2011, four years before Prime Suspect leapt on us.

* * *

At the time, I was working on a documentary called *Inside Nature's Giants*, a series that explained how animals worked, and demonstrated their evolutionary past by filming anatomical dissections. As part of our research, we had networks of scientists, zookeepers, national park rangers, and animal rescuers ready to call us in the sad event of a big animal dying. It was a strange job. We lived on standby, ready to rush our team to film scientists sent to carry out necropsies on giraffes, elephants, giant squids, and polar bears. On the morning the whale beached, I received a call from the UK Cetacean Strandings Investigation Programme (known as the "CSI of the Sea"). They told me to get to Kent fast.

Though I'd raced the couple of hours from my home in London, the whale had died while I was en route. It was a young but full-sized bull. The waters of the North Sea and the Channel are not good places for a sperm whale—full of shipping and industry, and not deep enough to hold many of the squid they feed on. Pegwell Bay is a large sandy beach that leans out into the English Channel. The spot chosen by Julius Caesar to land his triremes and launch his invasion of Britain almost exactly two thousand years earlier, it is easy to bring a boat aground there—but it is also hard to escape. The sperm whale had been spotted thrashing in the shallow water. Whales are not designed for gravity: They cannot support their own weight on land, and despite the best efforts of rescue teams, beached whales rarely survive. They can crush themselves against the ground, damage their internal organs, and become dehydrated. Toxic metabolic by-products build up from lack of movement and pool in their tissues. As the tide receded from Pegwell Bay, it left the body of the whale. Beachgoers who had chanced upon it gathered around, as people always have with beached whales. Some were dumbstruck and others cried. Some climbed on it while others touched its huge peg-teeth, and dogs nibbled on its blubber. The tide had placed it on its side. Blood ran from grazes where its thick but sensitive black skin and delicate gums had rubbed on the sand in its death throes. I touched its head, which was warm in the cold air.

Over the course of the day some forty people assembled: teams of scientists and their volunteer helpers in oilskins, a film crew of ten decked out in bright orange waterproof overalls, workmen in high-vis jackets, police in their dark blue. Our only hope of moving a forty-ton animal and dissecting it was via the medieval armory displayed in front of the creature: an assortment of hooks and flensing knives, specialized blades and straps. We rented heavy machinery to be driven down onto the beach. It got dark at half past five in the evening, so generators were fired up to power arc lamps on telescoping cranes, and their white light shone down on the whale. At one end of the body sat a 360-degree excavator and at the other a backhoe. Worried that this wasn't enough, we'd roped local tree surgeons in, and they arrived clasping diamond-toothed chainsaws with a mixture of confidence and disquiet.

To get inside any whale is hard work, but a sperm whale (*Physeter macrocephalus*)—the largest of the toothed whales—is a deep-sea diver, built to withstand pressures that would crumple your organs, crush your skull, and fold your ribs into a pizza tray. The police had agreed to give us two tide cycles—about twenty-four hours—to dissect and film the whale. In exchange, we would help arrange to cut it into smaller pieces and pull it off the beach so that it could be buried by the council. Disposing of a dead whale is a major hazard. If you dig a hole on a beach and cover it in wet sand, the whale can work its way back to the surface; whereas if pulled out to sea, it can endanger shipping and end up floating back to shore. They can even explode in transit: One whale being carted on a flatbed through a Taiwanese town did so, covering vehicles and shopfronts in viscera in the process. Some authorities try to preempt this and dynamite the whales, but this can backfire, too—sometimes spectacularly, as in Florence, Oregon, in 1970, when a detonated sperm whale rained giant chunks of blubber as far as three hundred yards away, flattening nearby cars and only narrowly missing gawping bystanders.

Joy had flown in at two o'clock that morning. Despite her lack of sleep, she was fired up and making sure everyone knew their jobs. She directed the team to make a series of small incisions around a section of the whale's belly. Gas trapped inside hissed out, a vital release to ease the pressure and prevent an explosion. Slowly they sliced their way through the smooth gray-black outer skin, before cutting first through the blubber and then the corset of fibrous connective tissue that wraps around the muscles of the whale.

Whale detonation, Florence, Oregon, 1970.

As Joy sliced though this, it made a crackling sound, the strands pinging beneath her knife like hundreds of taut elastic bands. Her team gradually made a huge sideways U-shaped incision, running up the belly. At the base of the U, they cut a hole and threaded a great rope through it, which they passed into the teeth of the bucket of the backhoe.

Everyone stood clear. As the arm of the machine pulled with a great rending sound, a colossal tab of flesh, the size of two king-sized beds, peeled off, revealing the whole meaty inner abdomen. We were into the muscle now, the whale's six-pack. The meat was black-red, the color coming from the muscle's enormous concentration of myoglobin proteins, which trap oxygen just as hemoglobin does in blood cells. The whale uses its flesh as a scuba tank, slowly releasing oxygen from its muscles during ninety-minute dives. Joy needed to get to the guts first—they would help tell us how healthy the animal was, what it had eaten, and if it had parasites.

She was working her way carefully through the muscle wall when her knife slipped in a bit too far, puncturing an intestine beneath. It was as if a shotgun had been fired out of the whale, a roar of steam and gore shooting out and across Joy's face. She was wearing safety goggles but otherwise was covered. "Wipe, please," she said. "Has anybody got a wipe?" I looked at Jasmine, our sound recordist. The "fluffy dog" windshield covering her boom microphone had been in the line of fire and now trailed grayish stringy goop. Her boots were sunk into a foot of blood and sand and intestinal fluids. Anna, the assistant producer, stepped forward with a wipe and gingerly cleaned the whale guts off Joy's face.

For most of history, dead whales have been our main source of cetacean information, and the whale hunters our experts. Whalers, not naturalists, named them: The "right whale" was considered the right one to hunt because their bodies would handily float once killed; the Bryde's whale named after Johan Bryde, a Norwegian who built a huge whaling station. Whalers, not anatomists, described their parts. For instance, the structure called the "junk," the lower part of the sperm whale's nose, which is part of its sophisticated communication anatomy, is so named simply because it was less valuable to whalers than the rest. (Imagine if the trunk of an elephant was called a "rubbish," or the feathers of an eagle the "inedible.")

That whales and other marine mammals possessed voices had been suspected by some whalers, who had noticed that right at the moment they harpooned one whale, others, including those far away, would instantly take fright, leaping into the air and changing their behavior. The whalers had assumed that the struck whales were crying out. Captain William H. Kelley of the *Eliza* described to *Outing* magazine in 1890 how he'd once put his ear to the line leading from a harpooned right whale to the boat and heard a "deep, heavy, agonizing groan, like that of a person in pain."

In the 1950s, the biologist Malcolm Clarke stowed aboard British whaling ships hunting in the Antarctic. As the men hauled the steaming whales onto the deck and began the mechanized business of processing them, he would weave among them, dodging hooks and flying chains. He was fascinated by the guts of sperm whales, not only for what they could tell us about their owners, but as portals into the deep sea, where, back then, no human had been. He discovered parasites fifty feet long, and chunks of ambergris, the fragrant orange waxy buildups of whales' digestive juices, like gunky meteorites worth their weight in gold to the perfume industry for their unique chemical characteristics (and still used in Chanel No. 5, among other classic scents). He also found beaks. Within one sperm whale stomach alone he counted eighteen thousand of them. The beaks belonged to squid, whose mouth parts are the only indigestible bits of their bodies (excepting monsters like the giant squid, whose suckers have "teeth," and the colossal squid, whose tentacles sprout rows of hooks). From these remains alone he discovered new species of squid, and through his scientific reports we began to realize the scale of the battles taking place between sperm whales and their huge molluscan prey in the dark, cold, crushing depths.

On the beach in Kent, I traced the ringed scars on the sperm whale's body, where the tentacle sucker teeth of the giant squid he hunted had lacerated him, presumably before being sucked into his maw. It was astonishing to think of what this whale might have experienced. He had perceived mountain ranges, valleys, life-forms, and chemical systems that are the stuff of science fiction; his dives had likely skirted deep-sea vents that spew sulphurous clouds of white fluid at 400°C, swimming above mountain ranges bigger than the largest on land, past sharks that live for more than four hundred years. It is as though he had lived on a different planet. If he hadn't died, he could have seen seventy years or more of deep-sea exploration—an estimate, as we are not sure how long sperm whales live, nor any other whales, for that matter.

What killed this creature? There are so many reasons whales are increasingly washing up on our shores. They can get sick and injured naturally. Some are killed in battles with other whales and sea creatures. But some are so full of heavy metals they are treated as toxic waste. Others have huge balls of plastic in their stomachs. Others still have clearly been hit by boats, or fatally caught in fishing nets. Navy sonar and underwater industry can be fatal for cetaceans: To these super listeners, a sonar pulse close by could be like a sound bomb. Mass strandings of many different species of whales, dolphin, and porpoise—including some involving hundreds of animals—have been linked to Navy exercises. Some bodies demonstrate damage to their hearing systems. Others seem to have suffered the bends. A recent study found that beaked whales are rendered so terrified and disoriented by sonar at certain frequencies that their hearts malfunction and, incapacitated by pain, they wash ashore and die. They are literally scared to death.

The verdict of the necropsy for our whale was that he had got lost swimming south from the deep waters of the North Pole, gone the wrong way around Scotland, missed the deep, safer waters of the Atlantic, and ended up in the shallow North Sea, where there was no food and much human disturbance. Disoriented and weak, he'd beached and died.

Under Joy's lead, the team spent several hours digging into the whale's empty guts, gradually removing them from the body and driving them away in the buckets of the machines. The knife-wielding teams then encountered the lungs, springy and snug beneath the animal's colossal ribs. As a sperm whale dives, its lungs are crushed by the pressure, the air inside compressed. In these conditions, a human's ribs would crack, but sperm whales' ribs have evolved hinges, so they neatly fold away. Joy showed us the fluid that lubricated the joints of these skeletal concertinas. It was hot now inside the whale, a plume of white steaming out into the wintry air. "Feel there, to the left, that's it," Joy said, as she plunged my hand around the whale's lungs to something firmer, its surface dark and shiny. It was the whale's heart.

Joy is five feet tall and fearless. As a young woman, she'd dreamed of being a jockey until her father told her it was no job for women. Now, she rode the innards of the whale, legs clamping to anything vaguely firm in the mashed purple blancmange of its body cavity, using a knife to get purchase on the treacherous tissues like an ice ax. Wrestling a lung to one side, Joy climbed inside the whale and sat fully in the cavity we'd excavated, a set of ribs extending above her, her feet and lower legs disappearing into a swamp of guts and blood. She removed the heart with a giant hook. It was the size of a desk.

As she cut, Joy explained how the whale had understood his world. Cetacean senses are different from human senses. Their smell and taste are mostly dulled, their vision is worse in general. But they can sense things we cannot—some cetacean species could even be sensitive to magnetic fields. Like you and me, they had imperatives; they must find mates, navigate around the ocean, and find food. Unlike humans, they have to do so in the dark. But because water is a denser medium, sound travels more than four times faster than it does in air; this gives opportunities for the keen-eared.

For sperm whales, and many others, sound is key. They listen their way through the depths.

Joy about to cut into the nostril to expose the "monkey lips," with author watching (holding black gloves).

We followed Joy as she toured us through the animal. I took it all in, as if being shown around a property by an estate agent. This whale's great heart, his powerful fluke, folding ribs, retractable penis, springy lungs, dark black muscles, endless squid-powered guts: These were all huge, because the whale was huge. But after all of this, what left the biggest impression was the largest part of his body—his head. And this was mostly made up of his nose.

Joy loves the snouts and heads of whales, viewing them as complex puzzles of anatomy, labyrinths of tubes and vessels often unlike anything in us, evolved for extreme conditions. The sperm whale, she said, had a snout evolved for sound, perception, and communication. It is impossible for a human to imagine what it is like to be able to sense sound like a whale or dolphin because we describe the world in our primary sense, which is vision. On the other hand, the sperm whale (like all its toothed whale and dolphin cousins) uses sound both to communicate via vocalizations *and* to sense its surroundings by firing sonic clicks ahead of it and listening to the returning echoes. When we describe how whales use sound to understand the world, we say they "see in sound," their "fundamental sensory and communication channel." We humans are not too shabby at figuring out where sound is coming from and at perceiving its volume and pitch, though we can't move our outer ears around like some other land mammals to really focus on a sound and pin down its origin. Yet we are miles off the abilities of the whales and dolphins, despite the fact that they have no visible ears.

Joy noted that life underwater has smoothed off cetaceans' fleshy external ear parts (pinnae), and instead sound is funneled into their internal ear parts through special fatty structures in their long jawbones. She sawed into the whale's lower mandible and explained how it could pick up vibrations like a satellite antenna. The sound waves are transmitted through the goo within the bone, up into the whale's inner ear, and onward to the brain, which interprets the scattered vibrations and turns them into a three-dimensional picture of the objects ahead: their hardness, shape, and density. These inner ears are likely more sophisticated than those in humans. Scans and dissections performed by Joy's colleagues of the inner ears of dolphins found they had thousands more of the receptive hairs used to sense sounds, connecting to double the number of auditory nerves. This has led scientists to conclude cetacean ears are "wired" for more complex ways of hearing and understanding sounds than ours. Not only are their ears better than ours, researchers working with bottlenose dolphins found their sonar is *better than any machine we have yet built.*

I was not surprised to discover that cetaceans had great hearing, given that they live in a medium that blocks light and transmits sound. But I was surprised to learn that they, unlike most good listeners, were also chatterboxes. I've sometimes tried to speak underwater and noticed it doesn't work very well. So, I asked Joy, how do whales speak?

I watched Joy's eyes light up. Twinned with their exceptional hearing were precise and powerful sound generators, she said. In fact, sperm whales make the loudest noises of *any* living creature. They do so using a set of lips set under the end of their single nostril right at the very front of the whale—its blowhole. The nostril of the whale on the beach was closed, as it would have been when the whale was underwater, as a submarine seals its tower, so no air escapes or water rushes in. This is strange for us to imagine—if someone claps a hand over your mouth so that air cannot escape, you are nearly muted. Try to make sounds with your mouth closed and your nostrils pinched shut. But this is how whales speak: The sound is generated by air moving around *inside them*, specifically through special passages within their heads. Joy led me to the front of the sperm whale to the source of his voice. I could see the cleft at the top right front of the whale where his single nostril was clamped shut. Around it were saw marks where the team's tree surgeons had tried to get in and given up; the skin and blubber were so tough here that their diamond-tipped saws had soon blunted. Joy got to work instead with her knives, sharpening and resharpening them as she slowly cut the nostril back. It was hard, exhausting work and took her more than an hour, but finally she was able to pull away the outer flesh of the top of the nostril and see underneath it. There, nestled in a cavity the size of a human head, were two fat, dark lips. They looked like two halves of a coconut pressed together. Joy explained they were phonic lips, more informally known as "monkey lips." They did look like the lips of a cartoon monkey, only hiding in a giant nostril. These lips were the origin of the most powerful and penetrating nasal voice in all creation.

When air passes through them, they vibrate against each other. This is the voice of the sperm whale. Think of it as a DJ's decks—useless unless connected to a speaker. The entire front third of the whale from snout to skull was its amplifier, a truck-sized sound system made out of a highly evolved nose. It would have taken days for her to dissect it all, and we were running out of time. To give a hint of it, Joy cut away a window along the side of the whale's head. Under the black skin was a mesh of overlapping white fibrous tendons; within this is a part of the "amplifier" called the spermaceti organ, which sits behind the monkey lips at the front of the whale and can extend back through more than 40 percent of its body length. As Joy cut through the ribbons, a stream of whitish, viscous fluid leaked out. She scraped it up onto her knife and held it out, and the liquid immediately hardened into a waxy stalactite.

This was the spermaceti. It's now thought vital in the transmission and perhaps focusing of the sound waves made by the monkey lips. But the first westerners to hunt the whales assumed this was the whale's semen, hence sperm whale, and the name "spermaceti," meaning "seed of the whale." A

Joy cuts through the case into the spermaceti, which flows out and hardens into wax in the cold air. You can see the rings of light scars across the dark skin of the whale—these wounds are made by the "sucker teeth" of its prey, the giant squid.

highly sought-after product, burning without smoke, and moving from solid to liquid state at precise temperatures, spermaceti lit and lubricated much of the industrialized world. In 1839, the whaler and naturalist Thomas Beale wrote that the sperm whale was "one of the most noiseless of marine animals." In fact, mariners had long heard the clicks of the whales but ascribed the bashing noises that resonated through the hulls of their boats to a hypothetical "carpenter fish." As with so many of our assertions about what animals can and cannot do, being very sure of something tends to come right before discovering you're very wrong. As I stared at the spermaceti in Joy's hand, a strange irony occurred to me: how the substance that helps give the sperm whale its voice, the loudest in nature, led to so many voices being silenced forever. I found myself tuning into Joy's voice, thinking about how her New York accent was formed, watching her breathe as she spoke with her characteristic exhalations and huffs whenever she talked about any particularly wondrous piece of whale machinery.

At the time I was focused on the business of filming: We had to record how this giant beast worked, from skin to innards, front to back, and we were exhausted. Filming the dissections of elephants, crocodiles, giraffes, and tigers, I had learned that you could tell a lot about an animal's priorities from its body. And this was a body in which a large part—perhaps more than a quarter—was used for the production and reception of sound. I had never seen anything like it.

Joy walked down the beach alongside the colossal spermaceti organ, demonstrating the path the sound would take as it resonated backward and within the whale from the monkey lips to where it would hit the skull. This was shaped like a satellite dish; by waving her arms, she demonstrated how, when the vibrations hit, they bounce off and tremor back through the lower part of the whale's colossal head, through the junk—which is really a sophisticated series of "lenses": oils, fats, muscle, and other tissues. These modify and focus the vibrations as they pass back and forth within the whale's head, channeling the noise out into the dark water as an astonishingly powerful and finely controlled click. As Joy paced this journey for me, she explained that the sound from a sperm whale could be up to 230 decibels. This is louder than a jet engine. In air, eardrums rupture at 150dB. Other cetaceans are also powerfully equipped. Scientists found that a dolphin pulsing underwater next to you can be more sonically intense than a rifle being fired at the same distance. The monkey lips, spermaceti, junk, and other mysterious structures in the sperm whale's head allow it to finesse the sounds it produces. Recent research into sperm whales has discovered that, as well as their highly directional echolocation clicks, which they use to interrogate the ocean, they produce a great variety of other noises: slow clicks, fast clicks, buzzes, trumpets, creaks, and "codas." A coda is a sequence of clicks. They send these out in bursts in many directions. The pattern of the clicks and gaps is thought to carry information in a sort of Morse code fashion. In one sperm whale community studied, over seventy different coda types were found. These sounds are thought to be the glue that holds their cooperative lives together—vital in keeping close, hunting, navigating, mating, and protecting one another. All in a dangerous, vast world without light.

This fits into a general picture of whales and dolphins as master sonic manipulators; indeed, cetaceans "use the broadest range of acoustic channels" of any mammal group.

After Joy had shown us how the dead sperm whale made its click, the animal's insides had been excavated to the degree that it was light enough for the machines to begin to move it, and the police asked that we let them take the whale away to be buried before the tide swept back in. As the machines tugged the whale across the slick mudflats Joy cried out for them to stop. She ran around to its penis, which had been pushed out from its normal streamlined position inside the whale. She crouched down and cradled its five-foot length in her arms. It resembled a giant black leech. She demonstrated how it was different from the human penis—it was fibrous and elastic, not erectile. Muscles at its base meant it was mobile and, similarly to your tongue, could curl and move around, penetrating in any direction. This was vital in zero gravity sex if you haven't got hands to hold on to your partner. "It is the most ultra-humongous penis in the animal kingdom," she said, her eyes blazing with admiration. Then she stepped back and let the diggers pull the whale away.

Joy, with the whale penis. Her goggles misted up from the heat.

Joy hadn't slept for more than a few hours over the previous three days. Her bright orange overalls were slick with dark blood, streaks of intestine, gobbets of black skin. Three times she'd had to have gouts of gut juice and other visceral explosions wiped from her face and tongue. As she watched the animal pulled away down the beach, she gave a long sigh, her breath misting with a plume like a whale at sea.

Back at home in London the next day, I slowly unpacked my gear. Everything smelled like whale, which whiffs like a fishy soup cooked in oil and then left in a garden shed for a winter. A lump of sperm whale meat fell from a wrinkle in my coat onto the floor. My cat, Cleo, licked and nibbled at it with enthusiasm. I thought of how crudely, with chainsaws and great hooks, we had probed the sophisticated and still largely unknown sensory structures within the animal. How ineptly we had explored an organ that itself had evolved to explore. As he had swum through unfamiliar waters on his last journey, he might also have heard the voices of other whales, his cetacean cousins—the zips and trills of hunting killer whales, distant pods of chirruping belugas, the lonely, strange tones of beaked diving whales.

A few weeks after the dissection on the beach, I traveled to the Azores, a chain of volcanic islands in the mid-Atlantic, to film living sperm whales. We spotted them in the distance, rising from their dives and exhaling with the distinctive single-angled spout that comes from breathing out from one nostril up on the top right of your head. Our guide maneuvered us into position near a whale without disturbing it. We slid into the water, and I was struck with the strange vertigo of the open ocean. With nothing to see in any direction, I felt like an insignificant floating speck above three miles of deep dark blue. Suddenly, I felt the clicks. I couldn't see the whale anywhere, and I didn't hear it, but I felt it—"khuck...khuck"—crisp, loud cracks in the air of my lungs and throat and sinuses. And then I saw a shape in the gloom move off.

I often think of what that feeling was: Was I being scanned? Animals with sight "see" other animals inasmuch as their eyes capture photons that bounce off them. Similarly, the whale would have received reflected sound from my body in its ears and "seen" me that way. But sound travels through things, too, and the echo of me would indicate not just the surface of my body but also my density. Could this whale have seen inside me? Seen me as no one had since I had been ultrasounded in my mother's womb? Years later, I wondered something else: Could I have been spoken to?

Sperm whales live in pods—close-knit family groups of fifteen to twenty animals, composed mainly of females and their young, with males roaming between the pods and often going solo. These leviathans operate nurseries; mother sperm whales will leave their calves to be communally guarded and even nursed by other whales while they dive deep to hunt for giant squid. When threatened, they team up to ward off predators, forming a circle with their heads on the inside and their tails facing outward as weapons. Young, vulnerable, and injured whales are protected within this "marguerite." According to sperm whale biologist Luke Rendell, there is evidence sperm whales may even care for other adult whales, providing food for those less able to hunt. A recent study he was involved in discovered that the whales appeared to learn how to avoid whaling ships and transmit this information to each other.

Very social whales like sperm whales are also the cetaceans most likely to mass-strand. This is where many seemingly healthy animals, sometimes many hundreds, become fatally beached together on the seashore. There are many theories as to why, from naval sonar to marine topography, but the intense bonds between the animals is thought a key reason. With smaller whales like pilot whales and dolphins, human rescuers are sometimes able to refloat them, but then watch heartbroken as the freed cetaceans swim straight back to their still-trapped fellows and their doom. Whales in these situations often vocalize intensely, and it's thought they are pulled back by their peers' anguished calls. They bow-ride together and die together.

Sperm whale societies are both held together and differentiated by their communications. They are very social and communicate frequently when they are near the ocean surface and setting out on their dives, exchanging codas with one another in turn in duet-like sequences. Living within each ocean basin there can be thousands of sperm whales, but it turns out they don't all "speak" the same way. Researchers listening to the whales have discovered that there are different populations, each of which has its own "dialect" of coda click-patterns. Scientists have dubbed these "vocal clans." It astonished me to learn that two whales of different vocal clans wouldn't just "speak" differently, they would live differently as well—whales of different vocal clans use different hunting techniques and hunt different prey, they care for their young differently and pass down other behavioral traditions unique to their clan. Even though their ranges overlap, the sperm whale vocal clans don't seem to mingle much with each other—they live as different tribes in the same seas, split by behaviors and perhaps unable to communicate with whales of other clans.

It's thought that the point when humans started creating social boundaries from one another based on different cultures was a critical moment in our evolution into an animal with large-scale societies based on cooperation. Speaking the same way as someone else helps you know who to trust and who to help. The people who study sperm whales and other highly social cetacean species like killer whales have observed them cooperating in the sea, hanging out in groups that do things a special way, ignoring or avoiding other whales who sound different. The power of these whale societies lies in their learned behaviors and pooled knowledge, and their ability to transmit this to one another. The scientists have found no better term for this than that these whales have "cultures."

When I think of whales passing information to one another in their cultures, I wonder what they might say and how long those whale cultures might have existed. I think of whaling, and how even if some whales survived, and populations rebounded, what of their cultures is now gone? I'm reminded of the British colonists arriving in Australia and seeing the indigenous people, who had no writing, and writing their cultures off—despite their spoken histories, which had been passed down for thousands of years, since before histories of Britain began. Since the cultures those colonists encountered didn't imitate their own, they were invisible to them and terribly damaged by their actions. The cultures of both whales and people are fragile; they can be lost.

I wanted to know what more we could deduce about these "cultural" animals from their bodies. Could their anatomy reveal whether communicating with them was a possibility, or fantasy? Two years after the humpback breached on me, six years after that dissection on the beach, I got my chance. Joy got in touch again to say that a very rare opportunity had come up: to examine the processor linked up to these sensitive ears and refined voices. She was offering me the chance to look inside the brain of a whale.

5

"Some Sort of Stupid, Big Fish"

atoms with consciousness...
matter with curiosity.

Stands at the sea...
wonders at wondering...
—Richard P. Feynman

Brains are complex and delicate organs—and whale brains especially so. Few whales are in good condition when they beach. Fewer still are reached in time to extract the brain before it decomposes. It's the first organ to go because the sensitive tissues are pressure-cooked deep inside the dying animal's skull by body heat the whale cannot release. And rare are the people with the skills to extract and preserve them. Cetaceans were long thought to have simple, undeveloped brains, because whenever scientists would get inside a dead dolphin's head, it had often already turned to sloppy mush. A good-quality whale brain is gold dust.

In order to obtain a whale brain for examination, the stars have to align: The whale must be freshly bigger than most industrial freezers, which aren't easy to drive to the sea and pop a whale head into, this doesn't happen often. I had long given up hope of ever seeing such a thing. But in 2018, Joy called me to say she had two on the way. She had been given the chance to dissect a stillborn baby sperm whale, as well as the head of a young minke whale—which is a kind of baleen whale, like a thinner, smaller humpback.

Both had been recovered some time before, and stored deep in the Smithsonian Institution's freezers. A refrigerated truck was to drive them the couple of hundred miles to New York, where Joy and her neuroanatomist colleague, Professor Patrick Hof, would be waiting at their lab at the Icahn School of Medicine at Mount Sinai in Manhattan. And if I wanted, I could come and peer into a cetacean mind.

Joy invited me and my film crew to sleep at her house in the suburbs, which she shared with her husband, Bruce, a bespectacled man with a neat beard, himself an accomplished medical doctor and scientist. The doctors welcomed us with exuberance, like learned Ewoks. Their house overlooked a creek where they'd go kayaking. Their basement was filled with whale paraphernalia—the eye of a humpback in preserving fluid, tables laid out with baleen and bones. It struck me that Joy was a marine mammal hoarder. Upstairs, she showed us her pet mouse, Spinelli. She'd learned that I had some Jewish heritage, and she and Bruce had laid out a suitable feast: matzo ball chicken soup, bagels of every description, plate after plate of delicious food. After dinner, she and Bruce played us a duet on the guitar, singing love songs from the sixties together. Later that night, in bed, I thought of the whale's eye in the basement beneath where I lay.

The next morning, we met Joy at the hospital staff car park at four o'clock, our little team clasping our coffees for dear life, much as people hold bear spray when they hear a grunt in the woods. Joy had started even earlier than us but showed no sign of fatigue. She and her colleague Patrick Hof both teach at the medical school and are also employed at the affiliated hospital, where they study how human anatomy relates to other species. We rode the elevator up to their floor, past rooms for patients, waiting rooms for relatives, consulting rooms, and teaching rooms, and found ourselves in a storage space surrounded by marine mammal skeletons, the skulls of killer whales filled with huge slashing teeth, and a line of grinning sea lion skulls. Joy warned us not to walk into the next room, which was full of dead people. In Joy and Patrick's domain, rooms for human dissection and the teaching of anatomy do double duty for dolphins, and in the depths of the hospital, powerful machines for investigating human brains are used for exploring cetacean anatomy, too. With Joy's help, Patrick has built up one of the world's most extensive marine mammal brain collections, with about seven hundred specimens of sixty kinds of whales and dolphins. Out the window, the orange light of the morning reflected off the high-rises around us, and joggers chugged through Central Park beneath. The smell from the cadavers was sweet and almost pleasant, until you remembered what it was.

Joy and Patrick use the hospital's advanced scanning machines—MRI and CT scanners—to take 3D pictures inside the heads of dead whales without having to cut them open and risk ruining the brains. Some scientists have even managed to scan the brains of living dolphins, showing them "lighting up" as their brains worked (likely wondering what the hell was going on).

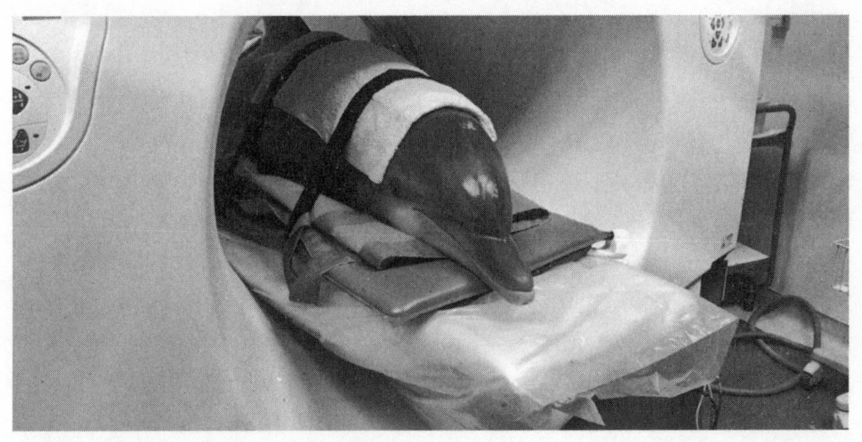

Here a dolphin undergoing rehabilitation has its brain scanned. (Activity conducted under a Stranding Agreement between NMFS and TMMSN under the MMPA.)

There are many scans of dolphin brains but very few of whale brains. This makes sense given that the biggest hospital scanners can scarcely accommodate a very large human, let alone an animal the size of a small hospital ward. Fitting an adult humpback into an MRI would be like trying to get a melon through the hole of a bagel. The two baby whales Joy had procured were just small enough to fit.

Access to Mount Sinai's scanners during the daytime was for patients only, and dead whales had to come in before the scanning department opened to the public. Seeing the decapitated, frozen head of a minke whale being carted past might be disconcerting for patients, so the whales were wrapped in plastic sheets on their gurneys. As we wheeled through various brightly lit corridors, down service elevators, past waiting rooms and sleepy patients walking alongside their IVs, none suspected our strange cargo—at most, a passing speed-walking doctor turned to look for the source of the strangely marine smell. In the MRI suite, there was a door with many warning signs and a window latticed with fine wire mesh. In the next room was the Magnetic Resonance Imaging machine (MRI), which resembles a giant white doughnut with a platform that patients (or baby whale heads) could be placed on, and gently moved through the machine. Joy told the technician, Jonny, that he was the first person ever to scan a sperm whale in an MRI.

The atmosphere was quiet as the team grunted and lifted the tiny sperm whale, the first to be scanned, onto the platform. Its dark skin was damp and cool. Patrick moved laser crosses along it to align the sensors as the machine whirred to life. Inch by inch, the baby whale progressed through the scanner, its head itself a huge and powerful scanning machine capable of discerning the densities of different tissues. For two hours the whale heads were scanned and turned. As they heated up, their juices dripped onto the platform and the floor. Then it was time for humans to reclaim their hospital, and the juices were mopped, the data saved, and the whales wheeled away.

Upstairs, Joy and Patrick had no time to lose. The brains were defrosting fast and had to be removed from the animals' skulls in the next few hours. In a room the size of two badminton courts, one end full of two dozen human cadavers, I watched as Patrick, a competitive épée fencer, and a mean hand with a scalpel, too, cut through the muscles and tissue around the back of the minke's skull. As the sun rose higher, the New York skyline brightened behind him as he used a saw to slice into the bone surrounding the brain, releasing a smell like burned hair. He cut a neat panel in the skull, like a burglar piercing the glass of a museum window, and teased the porridge-colored organ through the gap and into a jar of preservative fluid. Another whale brain to join its fellows in a vault, with all their pickled, unknowable thoughts.

The brain would be preserved and later dissected. Sometimes it would be cut into millimeter-thin slices stained to discover and trace the routes of individual nerves, other times into cruder sections. Or it would be kept intact to compare its shapes, grooves, and bulges with those of other specimens, including humans. Measuring and mapping, trying to see which structures resembled those in our brains, and which parts were totally different, Patrick and Joy made a good double act. It would take days to thoroughly examine the hugely complex scans. But Patrick brought up some of the images on his screen. Using a computer program, he could whiz through the whale's brain. It was mesmerizing to watch: the circle on his monitor like a porthole on a ship, the whorls and knots of brain being revealed as he sped through them, adjusting the controls to highlight blood vessels, denser tissue, connections, and convolutions. Although I was fascinated, it was difficult to identify the differences between the brain areas and tissue types that Patrick paused to point out. The Latin names for brain regions, one after the other, passed through my skull like CT scan rays, leaving little trace in my mind.

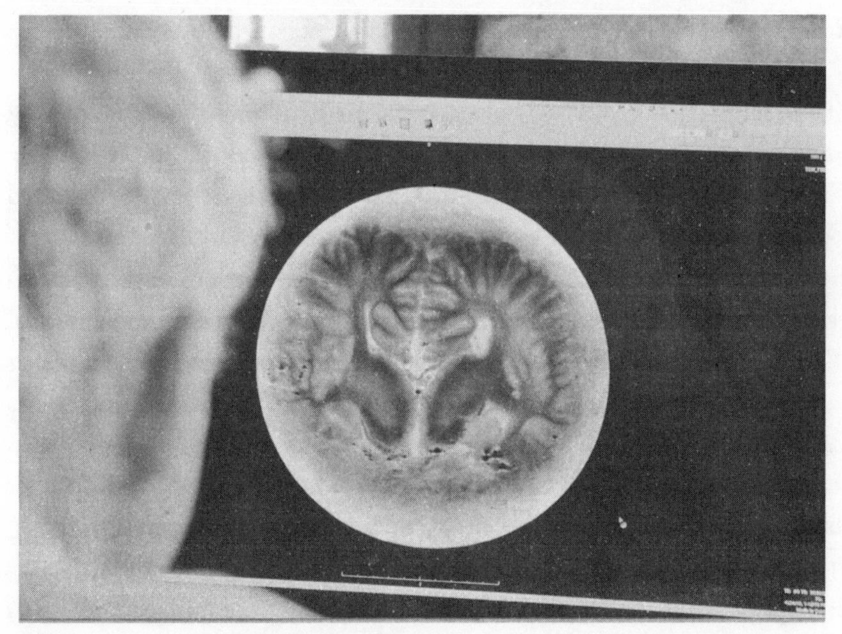

Patrick navigates through the brain scan.

* * *

I'd learned by this point that comparing brains is a difficult business in general. In explaining how clever we humans are, we often point out the extraordinarily large size of our thinking organs. Their bulk is the bane of childbirth and consumes 90 percent of the glucose in our blood. But size itself is not a clear guide for comparing animal intelligences, as some bigger animals with larger brains seem to lack the cognitive abilities of smaller ones. Size, as the saying goes, isn't everything. Relative brain-to-body size, how wrinkled and complex brains are, the thickness of their layers, the structures within them, and the types of neurons these are made of are all helpful—though our human brains are, naturally, the yardstick that other brains are measured against.

And yet it is impossible to look at a whale brain and not be surprised by its size. When Patrick first saw one, despite knowing they were big, its mass still shocked him. The human brain is about 1,350 grams (about three pounds), three times larger than our big-brained relative, the chimpanzee. A sperm whale's or killer whale's brain can be ten kilograms (twenty-two pounds)! These are the biggest brains on Earth and possibly the biggest brains ever, anywhere. It's perhaps not a fair comparison: In relation to the size of our bodies, our brains are bigger than those of whales. Ours are similar in proportion to our body mass, as the brains of some rodents; mice and men both invest a lot of themselves in their thinking organ. But we both lag far behind small birds and ants, which have much bigger brains compared to their body size than any big animals.

The outer layer of a mammal's brain is called the cerebral cortex. In cross-section, it looks a little like a wraparound bicycle helmet sitting on top of the other parts of the brain. This is the most recently evolved part of our brains, and it was by using their own cerebral cortexes that brain scientists have learned that this area is responsible for rational, conscious thought. It handles tasks like perceiving senses, thinking, deciding to move your body, figuring out how you relate to the space around you, and language. You are using yours now to read and think about this sentence. Many biologists define "intelligence" as something along the lines of *the mental and behavioral flexibility of an organism to solve problems and come up with novel solutions*. In humans, the cerebral cortex, acting with other bits of the brain (the basal ganglia, basal forebrain, and dorsal thalamus), appears to be the seat of this form of "intelligence." The more cortex you have and the more wrinkled it is, the more surface area is available for connections to be made—and voilà! More thinking. We humans have a really large neocortex surface area, but it's still just over half that of a common dolphin, and miles behind the sperm whale. Even if you divide the cortex area by the total weight of the brain to remove the cetacean size advantage, humans still lag behind dolphins and killer whales.

But there are other measurements in the cortex that seem to be associated with intelligence, and here, dolphins and whales lag behind humans. The more neurons are packed in, how closely and effectively they are wired, and how fast they transmit impulses are also extremely important in brain function. Just as the composition and layout of the chipset in your tiny, cheap cell phone allows it to pack more computing power than a 5.5-ton room-sized 1970s supercomputer. Both cetaceans and elephants, the biggest mammals on land and sea, seem to have large distances between their neurons and slower conduction speeds. In raw numbers of neurons, humans here, too, have the edge, with a human cortex containing an estimated 15 billion neurons. Given the larger size of cetacean brains, you'd think they'd have more, but in fact their cerebral cortex is thinner, and the neurons are fatter, taking up more room. Nevertheless, some cetaceans such as the false killer whale are close behind human levels with 10.5 billion cerebral neurons, about the same as an elephant. Chimps have 6.2 billion and gorillas 4.3 billion. Further complicating comparisons, whales have huge numbers of other kinds of cells, called glia, packing their cortexes. Until recently, we believed these glial cells to be an unthinking filler, but we've now discovered that they actually seem important for cognition, too. I don't know about you, but all this cortex measurement and comparison makes my own feeble organ hurt.

Patrick moved through the scans, the one hundredth marine mammal they had analyzed this way, zooming and measuring, exploring through symmetries and fractal patterns as if flying through a monochrome kaleidoscope. Questions spilled out of me: Could the brains tell us if whales or dolphins might have the capacity for consciousness? Could they allow these creatures to conceive of others? Patrick would not be drawn into discussing these matters. He felt that we simply did not know enough. Many others, however, have been more opinionated.

One study concluded that humans have five times the information-processing capacity of cetaceans, whom they placed beneath chimps, monkeys, and some birds. But in the same study, horses—with smaller brains than chimps—were found to have five times the number of cortical neurons. Does this mean horses are smarter than chimps? A major confounding factor in these types of comparisons appears to be that *every* factor is quite confounding. Estimating numbers of neurons is a very rough science, so the raw number comparisons are crude. There are lots of different kinds of neurons, and they are arranged in different configurations and proportions in different species. We know all these variations mean *something*, that they will determine what brains are capable of, but we don't know yet quite what, or how that might change from one moment to the next in different parts of the brain. There are a lot of assumptions at play, and it can be misleading to extrapolate from one brain to another.

This also applies to comparing cognitive ability. Trying to infer from brains and their structures which animals are "better" at cognition and ranking animal brains in order of "intelligence" is as treacherous as it is tempting. Professor Stan Kuczaj, who spent his lifetime studying the cognition and behavior of different animals, put it bluntly: "We suck at being able to validly measure intelligence in humans. We're even worse when we try to compare species." Intelligence is a slippery concept and perhaps unmeasurable. As mentioned earlier, many biologists conceive of it as an animal's ability to solve problems. But because different animals live in different environments with different problems, you can't really translate scores of how well their brains perform. A brain attribute is not simply "good" or "bad" for thinking, but rather varies depending on the situation and the thinking that brain needs to undertake. Intelligence is a moving target. What confounds this dilemma further is that individual animals within a species have varying cognitive abilities. To quote the Yosemite National Park ranger who, when asked why it was proving so hard to make a garbage bin that bears couldn't break into, said, "There is considerable overlap between the intelligence of the smartest bears and the dumbest tourists."

This comic by False Knees skewers the anthropocentric concept of intelligence beautifully.

We know little of the problems that the brains of cetaceans must contend with. They have evolved to process the challenges of very different lives—some solitary, some members of groups of hundreds, from giant hunters of the deep to tiny river dolphins. Faced with all these caveats and uncertainties, I began to see the wisdom in Patrick's hesitancy to infer too much from this terra incognita.

I had an odd thought as I had watched Patrick and Joy scan the whale brains. Perhaps from lack of sleep, I found myself imagining scanning their heads, stripping past the skin and muscles and bones, and looking at them as sense organs, eyeballs, ear canals, smell and taste receptors, floating in space and connecting back, via nerves, to the strangely bland organ, the hyperconnected fatty bolus where their thoughts and personalities and memories lived. If I looked at these floating brains, peering within, would I know them better? The human brain is often referred to as "the most complicated thing in the universe"—by scientists, spiritual leaders, and journalists alike. It is indeed a very complicated thing. But as whale brains also seemed, well, pretty damn fancy, I asked Patrick a simple question: Do whales have thoughts? He paused for a long moment. "Whether they have thoughts that are constructed in the same manner? Very possible. There's no reason that the same networks of nerves that served consciousness and memories in us cannot also exist the same way in whales."

Encouraged, I leapt ahead. Might whales think like us, then? With consciousness? Was there any indication they might have the brains to speak to each other like we do? "You know, there's potentially a lot of wishful thinking in all of this," he replied.

Wishful or not, Patrick had fueled a fair bit of this thinking himself. In 2006, he and his colleague Estel Van der Gucht published a paper in *The Anatomical Record* that set the brains of neuroscientists fizzing across the world. When examining preserved slices of human brain, he encountered an unusual-looking neuron. Instead of being shaped like a branch, cone, or star, it was long and thin, and very big. He realized he was seeing a von Economo neuron (VEN), a type of brain cell that was first described more than a century before but had been long ignored. These special nerves had been thought uniquely human. Then, in San Diego, his colleagues found them in the great apes (our close relatives the chimpanzee, gorilla, orangutan, and bonobo) but not in more distant relatives like lemurs. Patrick and others began to hunt for the cells, looking through the brains of more than a hundred species, but only a few seemed to have them: humans, the great apes, elephants, and cetaceans. We are distant relatives to elephants and whales, with our common ancestor evolving around the time the dinosaurs went extinct, over 60 million years ago.

Apes, elephants, and whales have much in common: We live a long time, are highly social, very intelligent, extremely communicative, and possess large brains. The VENs appeared to have evolved independently in these three groups, after our ancestors had split into different species, via convergent evolution, a process in which the pressures of natural selection lead to the same features developing in unrelated creatures.

The VENs seemed to be found only in certain areas of the human brain: the frontal insula and cingulate cortex. These regions are used when we feel pain, or notice that we've made a mistake, and when we feel things relating to others. A VEN lights up when we feel love, when a mother hears her baby cry, when someone attempts to ascertain another's intentions. In humans, the parts of the brain that relate to high-level cognitive functions, such as attention, intuition, and social awareness, are larger than in most other mammals. This is true for whales, too. And VENs are present in both species. As Patrick put it, "The cells that make human integrative experience quite unique are also present in large whales."

While we still don't know precisely what these cells do, there are some intriguing interpretations. In both whales and humans, the neocortex appears to have special "integrative centers" that process and integrate the information coming in from the sensory and motor areas. They chew over the signals they've received and communicate with one another in networks. This ability to integrate information from different brain regions is vital: It adds complexity to our perceptions and allows us to carry out advanced cognitive processes such as artistic creation, decision making, and language learning. Patrick and his coresearcher John Allman speculated that the VEN cells evolved in response to a *need*. To send signals quickly between their integrative centers, brains need highways, and VENs, according to Patrick, "are like the 'express trains' of the nervous system." Considering the functions of the regions that house these neurons, and the social nature of the species that have them, these high-speed brain links could be used when thinking about others—for empathy and social intelligence. Some are skeptical of this suggestion, believing that large, complex whale brains with VENs are simply necessary for coordinating enormous bodies in a 3D sea environment. Others say these impressive brains are required to process all the sophisticated information involved in echolocation: Their brains have evolved these structures because of *how* they sense, not because they are actually mulling over the results.

In 2014, Patrick and colleagues found VENs in more species than was previously thought, discovering the neurons or similar cells in the brains of cows, sheep, deer, horses, and pigs. This information was interpreted by some as evidence that VENs didn't herald any particularly impressive cognitive functions. To me, this story mirrors so many in biology. We discover something we think is unique to us. Then we find it in other animals and begin to question whether it is special anymore. But if you've spent time with cows and pigs, it's not surprising to think they might have neural hardware for thinking about others and social intelligence. This is all very recent information, and scientists like Patrick are explorers of a new frontier. It may turn out that a VEN in one beast might do something very different from a VEN in another, just as a piece of electrical cable can send both a signal to turn on a light bulb and a passionate email to your lover's computer. For Patrick, VENs are only a small piece in the sophisticated wiring diagram of the brains of some species, a diagram that is still very much in the process of being filled in. Discoveries, comparisons, hypotheses, and extrapolations connect and interweave, and will hopefully, eventually, build a clearer picture. We are at a frustrating moment; all this discovery without knowing what it means. In the words of one neuroscientist: "We don't even understand the brain of a worm." Perhaps that is simply a hazard that comes with poking around in the most complicated, gloopy mush in the universe.

Joy made a helpful comparison: If you were an alien explorer in the seas of Earth and you came across a bottlenose dolphin and a similar-sized shark, you might be puzzled. The animals live in the same sea, may hunt the same fish, and need to survive the same conditions, but the bottlenose has a far bigger brain. A brain that seems in many ways very similar in composition and structure to that of the highest mental achievers on the planet, and in other ways very different. Why would there be such a discrepancy between the dolphin and shark?

In 2007, Lori Marino, along with Joy, Patrick, and many other biologists, published a paper called "Cetaceans Have Complex Brains for Complex Cognition." They reached their conclusion by assessing all the current research, but also by looking back in time at the fossil record. Neurons and cortexes don't preserve well for millions of years, but skulls do, and skulls reveal brain size. Cetacean brains suddenly got bigger about 10 million years *after* they had already moved into the sea. This surprised some scientists who had previously linked cetacean brain evolution to adaptations to water and cold. Logically, any brain adaptation related to aquatic life would have happened sooner. The coauthors theorized that the leap in brain size took place as cetacean behavior became more complex, more social.

For many whales and dolphins, the challenges of life are impossible outside of a social group. To successfully live in a social group, to compete and cooperate, requires thinking you don't need to do as a loner. Patrick elaborated: "They communicate through huge song repertoires, recognize their own songs, and make up new ones. They also form coalitions to plan hunting strategies, teach these to younger individuals, and have evolved social networks similar to those of apes and humans." A social animal needs more brain hardware on which to run the software of culture.

I tried a final time: What could he tell about whales definitively from investigating their brains? Patrick said it was absolutely clear that whales were extremely intelligent, with impressive neural systems, components of which we had previously thought existed only in humans. Like so many scientists I've met who studied whales, Patrick would mention an exciting whale attribute—something relatable to human existence—and then immediately caution: We should not anthropomorphize. But he insisted that we could not consider whales completely inferior to ourselves: "There are many people who think they are sort of stupid, big fish, right?" he said. "And no, they are not, definitely not." Trying to figure out whether whales might think like us was both more complicated and more compelling than I had anticipated, every answered question a doorway into a further mystery.

It was late in the day now. The whales had been scanned and their brains saved—and everyone was exhausted. Patrick had medical students to teach and Joy whale faces to deflesh. I left the hospital and walked out into the streets of Manhattan, picking up on the moods of the people I passed from their gaits, overhearing their conversations, judging how to weave among them, avoiding the eyes of the strange man on the subway, laughing at a joke with a friend at dinner, feeling warm when I hugged them goodbye. I thought about the neurons firing within me, the brain centers integrating these sensations and thoughts. In the waters off New York City, just miles from where I stood, humpback, fin, and sei whales can be found. Did their brains also flash with complex thoughts, articulated by strange aquatic voices, heard by sensitive hidden ears?

I looked up dizzied at skyscrapers, choked on diesel smoke from thousands of idling engines, was dazzled by clothes in resplendent hues. No whale has ever made anything like this, I thought. But was I biased? Is it just human nature to believe we are more advanced than other species, using our accomplishments as evidence? When we think about which animals are clever, and relate them to ourselves, we instinctively point out our impact on the world—our tools and our constructions. Animals can't do these things as well as we can! Beavers build dams, but they can't write books. Orangutans make leaf umbrellas but not wheels. Insects construct cities but not libraries. *Look on our works, ye termites, and despair!* But there are other reasons a whale can't build a cathedral. It is physically harder to build a civilization in the sea, where nothing stays still. You can't light fires in water and create new compounds and structures; you can't store food outside your body; you can't use fins to manipulate tools. Would a marine *Homo sapiens* have been able to produce any of the wonders of the world, the signs of our civility—clothes, tools, buildings, agriculture, written records? It is doubtful. But a more straightforward reason for the lack of cetacean material culture could simply be that whales don't have minds capable of conceiving a cathedral, or a hammer to build one.

I had hoped that by looking into a whale's brain I'd know whether they might be clever enough to speak, but the waters seemed muddier than the Hudson flowing a few blocks from the hospital. One of the brains I'd seen inside Mount Sinai, a human one, had the capacity for language and

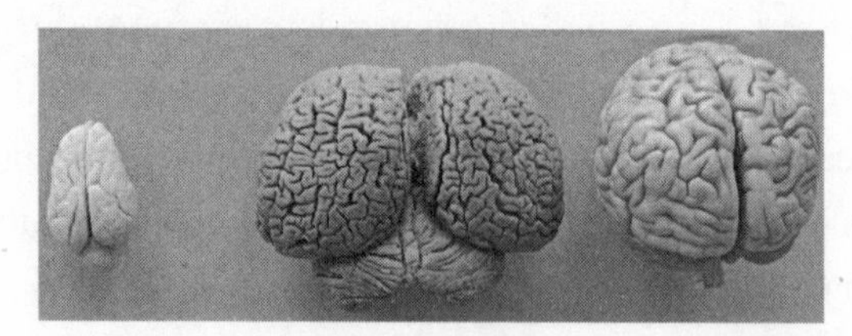

The brain of a bottlenose dolphin (center) flanked by that of a wild pig (left) and a human (right). Note the thicker gap between the hemispheres in the dolphin brain. This is thought related to something called "unihemispheric" sleep. Here one half of their brain rests at a time, allowing them to keep swimming up to breathe and one eye open to stay vigilant. I wish I could do this.

communication, for the appreciation of music, for the feeling of love, for the plotting of revenge. The other, that of a baby sperm whale, looked very similar. The whale brain and its scans didn't scream, "I'm an idiot," but neither did it scream, "I am Mozart."

Thanks to Joy and her colleagues, I now better understood the hardware, the parts of a whale that would be required to speak. Their powerful and sophisticated ears and voices hinted at the importance of listening and vocalizing in their lives. I had looked inside their brains and seen how the sizes, shapes, and constructions of these thinking organs suggested they possessed impressive capacities for cognition. Whales were clearly more than just large, stupid fish. But how much more? How much were the scientists I'd met extrapolating from these anatomical clues guilty of wishful thinking? I thought of the human cadavers I'd seen that day, faint shapes of people under sheets, laid out at the end of the room. Could you infer from looking at their throats that they had sung folk songs, or from their brains that poetry had made them cry?

Two years later, I met a man who told me a story that made me think back to my day with the brain scanners. His name was Duncan, and he was an underwater cameraman who lived in Bimini in the Bahamas with his shark-obsessed wife, Jillian. An incredibly laid-back presence, with a scruffy blond beard, Duncan has spent a lot of time in the water with large, sometimes scary animals, like oceanic whitetip sharks. He told me how he had been "scanned" many times in his life by dolphins and sperm whales, and he could feel it when they directed their sonar at him—he compared it to standing in front of the bass speakers at a loud concert. "You feel that vibration inside of your chest. That's kind of how it feels when an animal is scanning you."

One time, Duncan had been filming a pod of spotted dolphins using Kodak cinefilm. Each roll gave him eleven minutes of filming time underwater. The old mechanical camera was loud and made lots of clicking noises and the dolphins loved it—in his view, because "it probably sounded like a dolphin." Duncan had swum in with a pod led by a really old female, her age clear from the deep mottling of her skin. They'd headed toward the beach, and the dolphins seemed to be relaxing on the surface of the water among the floating sargassum. He started recording, but before long, Duncan heard the camera's magazine go *click!* He was out of film. The sun was setting, and as filming was over for the day, he lowered the camera to take in the scene. As he did so, the huge, old female, who was covered in battle scars, came directly toward him, "slowly," he said, "like a bus." She gently put her rostrum, or snout, onto his scuba mask. "Right here," Duncan said, pointing between his eyes. "Then she just started buzzing me, like, doing sonar." She held her position in the still water for a few minutes, and he held his, breathing calmly. He said it felt as if someone had shaken a can of soda and it was gently fizzing around his head—"a very pleasant sensation, to tell you the truth."

When Duncan told me this story, I thought back to Joy and Patrick scanning the baby whale brains, trying to sense what they were made of, what they represented, what they could do. I wondered what the dolphin—an animal with a built-in scanner, capable of interrogating the innards of living beings—could perceive in Duncan. What did she feel? What was she thinking about the human she beheld, if anything at all? I lingered on the picture, imagining minutes passing in the tranquil water, the evening light dancing across the dolphin as it held its rostrum to Duncan's face. A connection. Perhaps even a communication, of sorts.

For just one moment, that seemed like enough.

Atlantic spotted dolphins, Bimini, the Bahamas.

6

The Search for Animal Language

Man has great power of speech, but what he says is mostly vain and false; animals have little, but what they say is useful and true.
—*Leonardo da Vinci, Paris Manuscript F, fol. 96V*

My adventure at Mount Sinai Hospital had been strange, gory, and beautiful. I had seen inside a whale's brain. I had smelled it; I had touched it. With much of anatomy, you can easily infer the function of a body part from its structure. You can contract a muscle and see it pull on a tendon and move a bone. You can trace the path of a blood vessel, observe the tiny hairs within your ears vibrate, and the cells that turn these vibrations into electrical impulses. But in the brain, much of what you see is inscrutable connection, a tangle of wires.

After the visit, I took out Roger's second album of whale sounds, *Deep Voices (The Second Whale Record)*. Released on Capitol Records in the late seventies as a follow-up to his first album, it featured not just humpbacks but also blue and right whales. In some of the recordings, several different animals vocalize in short bursts with large gaps in between, like a "momentary squabble in their otherwise serene and gliding lives," Payne wrote in the liner notes. I listened to one called "Herd Noises." It sounded like buffalos disagreeing at sunset on top of slowed-down human voices, all roiling around one another for forty-three seconds. I listened again, and again.

How could one be sure this wasn't a language? And if it was, how on earth, with no translator, could you ever begin to try to understand what it meant?

A right whale cow and calf. Some mother whales are thought to "whisper" to their young while migrating, to keep predators from hearing their communications.

I imagined a whale sent from the ocean to study human communications. Perhaps she might make some recordings of us and note that we produce sounds between 85–255 hertz in bursts usually lasting from seconds to a few minutes and rarely more than an hour at a time. That these "words" are normally spoken between a few individuals in overlapping turn-taking, punctuated by other nonspeech noises like laughs or sighs or groans and claps of hands and stomps of feet. But how would this whale biologist know that these were words? Words that meant things, where order was important and gave the words greater meaning, with questions and answers? That those words represented abstract ideas and other people that weren't present, and even things that might not have happened yet or could not happen? How would she detect that we were speaking a language?

I already knew that the word "language" can get you into all kinds of trouble if you combine it with the word "animal." It's something that Roger has avoided by coining the term "whale-speak." But I had no idea of how truly upset it can make people. I once accidentally put four 150-foot lengths of handmade cloth bunting through a washing machine on a high-spin cycle and nearly drove myself to tears trying to untangle the sodden, intractable mess that emerged. If you're a fan of this sort of activity, I recommend researching the knotted ball of feuds and disagreements that is the cross-disciplinary discussion of what language is, why we have it, where it came from, and why it exists only in humans. Helpfully, many in the various fields seem to know the answers to all these questions. Unhelpfully, they seem to have come up with very different answers. It's a hellscape of nitpickery and grand theory, a world of big cheeses listing their absolute descriptions of what language is, if and where it lives in our brains, and why their rivals are wrong.

We are born a blank slate and acquire language like other behaviors through conditioning!

We are born with a special human Universal Grammar! A language instinct!

There is no Universal Grammar! But humans can build language from our cultures!

There is no "seat" of language in the brain, but a distributed "functional language system"!

Recursiveness is what makes our language special!

True language is only possible verbally, and only humans can learn to control our voices!

Language is a multifaceted phenomenon, grounded in but transcending individual cognition!

Even the definition of language changes within departmental fiefdoms and between one doyen's big publication and the next. One particularly weird disagreement in the history of linguistics was over whether American Sign Language (ASL) had enough prerequisite characteristics to qualify as a language, an argument you could explain to a deaf person using ASL. There continues to be no universally accepted definition of language. Of course, this could indicate a thriving area of research about something important but tricky that people are passionate about, or it could indicate that there are a lot of people with strong opinions about the matter and few ways of discerning whose is closest to reality. As I read up on these debates, perhaps the most consistent thing I noticed was the repeated assertion that language is a human-only thing, and that this claim was often made by humans who exclusively studied their own species. How could they know this for sure?

But mentioning the very idea of "animal language" to some scientists was like "waving a red cape in front of a bull." It seemed a surprisingly emotional topic. The primatologist Frans de Waal wrote that "the one historical constant in my field is that each time a claim of human uniqueness bites the dust, other claims quickly take its place." Could it be that as we discovered evidence that animals had the capacity for other things previously thought uniquely human—tool use, culture, theory of mind, emotionality, personality, perhaps even morality—the question of whether animals might have their own languages became more loaded? Was there something in human psychology that recoiled at the thought?

Another issue was that the word "language" meant different things to those who studied it and to everybody else. We weren't speaking the same language. If you told a person on the street that whales made sounds that communicated to other whales who they were, where they were, their emotional state, that they perhaps even gave other whales warnings or described elements of their world, that person would likely be satisfied that the whales were speaking *some* form of language. But for a biologist or linguist, these whales wouldn't be using "language," but rather they were vocalizing with their animal communication system.

So, what exactly is "language" for a biologist?

There are a lot of hurdles to answering this question pithily. One is that when humans communicate, we're often doing so in many different ways at once: not just using the words themselves, but the way we say them, and our body language. Think of the last time you told someone you loved them. Did you just say the words "I love you" in a flat monotone, face deadpan, slouched, eyes closed, arms by your sides? Or did you choose how you said them, with warmth, not too loud, not too quiet? As you spoke, what were your eyes saying, and your hands and your body? Did you angle yourself toward them, or turn away? Did you touch that person, or hold back? We forget about the other ways we talk when we talk. The scientific term for this is that communication is multimodal. Other animals also make use of multiple simultaneous communication channels. Which begs the question: If you're trying to unpick someone's communications, which signals should you choose?

Humans can communicate in many ways, but we cannot, at will, release combinations of pheromones from fifteen glands to summon and excite or warn or court our fellows, as the honeybee does. We can't peel back and display hidden neon feathers in a dance of electrifying winged semaphore, as can some birds of paradise, or change the color and reflectiveness of our skin in milliseconds like cuttlefish, with one side of our bodies displaying a come-hither to a suitor, while the other side flashes a warning to a rival.

Many human signals are unspectacular, as are our senses. Although we are pretty sensitive to some parts of the light spectrum, that range isn't terribly impressive; we are blind to infrared and ultraviolet wavelengths. Our ears cannot hear the sonic rumbles of elephant voices, as they are below twenty hertz, so their vibrations pass through our bodies with those of distant earthquakes. Nor can we hear the calls of bats and moths swooping past our windows at night, which are above twenty kilohertz. Cows have twice the hearing range we do; in fact, humans live in an audio bubble, deaf to the chatters of tarsiers, the call of the sloth, and even the complex trills of male mice. We can hear rats squeak, but not when they are happy; as when they become excited—such as when tickled—the pitch of their squeaks increases and is lost to us. Which means we only hear sad rats. Our skin cannot emit or sense electric charges, and we do not possess tiny lines of pits along our flanks to translate the disturbances made by animals moving near us into information. Rattlesnakes and starlings and elephants and hummingbirds and hammerhead sharks and electric eels and bluefin tuna have these tools in their sensory arsenals. When they need to communicate, they can utilize sounds we cannot hear, colors we cannot see, scents we cannot smell, and forces we cannot feel, and combine them with other signals. It is all lost on us.

This black-eyed Susan flower as it looks to us, on the left, and as it looks to a bee, who can also see in UV, on the right.

But it seems all these other animal communication channels are rather neglected because we humans love *words*! Our verbal language—sounds formed into words used in sentences in complex conversations, or written in squiggles on dead trees—is genuinely marvelous. It gives us, for instance, the power to invent abstract concepts and fictions and transmit them to one another. We hadn't detected capacities like this in the communications systems of other species, so many biologists decided that this human language was the ne plus ultra. It, and only it, was language. What, then, constitutes a human language?

In 1958, the linguist Charles Hockett published a linguistics textbook, which included a section called "Man's Place in Nature." Here he laid out a list of the seven properties of human language (which was later expanded and modified into a list of sixteen properties). The term "natural language" is often used here to differentiate between the languages humans use that have evolved without conscious planning—Mandarin or Spanish, for instance—and purposely planned and constructed languages like those we have created for machines, philosophy, logic, and Klingons. Hockett's list became known as the "design features" of language. Things like *semanticity* (words—the units we transmit—have meaning), *discreteness* (words must be transmitted in chunks with gaps between them), *productivity* (new words for things must be made up and used), *displacement* (the communications can transmit information about things happening somewhere else or in the past or future). If any animal communication system was to be credited as a proper natural language, they had to have all these traits. Hockett's properties together set our natural language apart from nonhuman communication systems and allowed us to compare between them. It isn't the only way of analyzing what makes a language, but it has been extremely influential.

Hockett already knew that some animals employed a number of his design features—bird communications utilized semanticity, honeybee dancing demonstrated displacement—but only humans had the full house, including some traits he didn't consider any animal to have, such as *cultural transmission* (learning your communication system from your peers) and *prevarication* (using language to withhold information or deceive). People immediately disagreed over whether all the features were universal to all human languages, and extra design features were added: Human language had rules about the order you could use words in (*grammar*), changes to what the combined words meant if you changed their order (*syntax*), and if you wanted to, you could stuff further clauses, adding layers of meaning, into your communications (*recursion*). Disagreements about what in particular constitutes language continue to this day, and their proponents' different perspectives have led to some "very bitter disputes." Meanwhile, some pioneering scientists felt that we still didn't have enough information to rule language out in other species. Perhaps it was there, but just invisible to us. So they set out to see what they could find, braving the slings and arrows of their "only-humans-have-language" colleagues.

They sought to discover not just whether animals had language abilities, but also to settle debates over where our own language abilities came from. Were they instinctive, or could they be learned? Were they physical or behavioral? One of the first places they hoped to find the answers to these questions was in our hairy cousins, the other great apes: chimpanzees, gorillas, orangutans, and bonobos.

Our close relations, these brainy, social apes—adept at tool use and all sorts of conniving—seemed ideal animal candidates to teach language skills. The great apes are like us, with similar anatomical signaling tools and sensory systems, and we could easily keep them in captivity and test them. There was lots of early success in communicating with apes using sign-based systems in which the chimps and gorillas could point at objects or tap screens, but whenever the researchers tried to teach them how to communicate using their voices like humans, they were frustrated. While primates were often skilled in mimicking human gestures or in making sounds in response to their human trainers, they just could not seem to articulate human words out loud—no matter how many bananas they were given. Although chimps have a wide vocabulary of grunts, pants, and hoots, they are thought to be in large part gestural communicators, using their hands, posture, and facial expressions to transmit information to one another. The gift of the gab—the manipulation of vibrating vocal cords, movement of tongue, aspiration of breath, and folding of mouth into the articulation of speech—seemed uniquely human among our close relatives. Why?

For a long time, it was thought that the reason chimps could not speak like humans was that their vocal tracts could not move like ours, and so they couldn't modify their vocalizations enough to form different vowel sounds. But this theory was upset by a scientist named Dr. William Tecumseh Sherman Fitch III, named for his great-grandfather, the U.S. Civil War general William Tecumseh Sherman, himself named after the great chief of the Shawnee people, Tecumseh. I had first met Tecumseh during a shoot for a film on the anatomy of big cats. He had asked me to fetch a vacuum cleaner, and when I returned with it, he set it to blow, not suck, inserted it into the trachea of a dead lion, and demonstrated how he could make it roar from beyond the grave. This ghost roar left a lasting impression on me. I bumped into him a few years later at the Vocal Interactivity in-and-between Humans, Animals, and Robots conference. A tall, broad-shouldered man in his late fifties, with a shaved head and a dark goatee, he has more than a passing resemblance to the renegade crystal meth–cooking chemist Walter White in *Breaking Bad*.

Tecumseh had been investigating speech in humans and other animals for more than thirty years. Since we'd last met, he'd been filming animals vocalizing, using machines that could see through their skin—such as live X-ray and CT scanners—which helped him observe what their sound-producing anatomy could do in real time. One of his subjects was a long-tailed macaque called Emiliano. Emiliano sat in a live X-ray machine and, well, monkeyed around—eating, vocalizing, lip smacking, yawning. This behavior meant Tecumseh and his colleagues at Princeton could make scans of the macaque's throat in all its range of movement, which he used to build a 3D simulation of Emiliano's vocal hardware, allowing Tecumseh to deduce the resulting range of sounds Emiliano should have been capable of making.

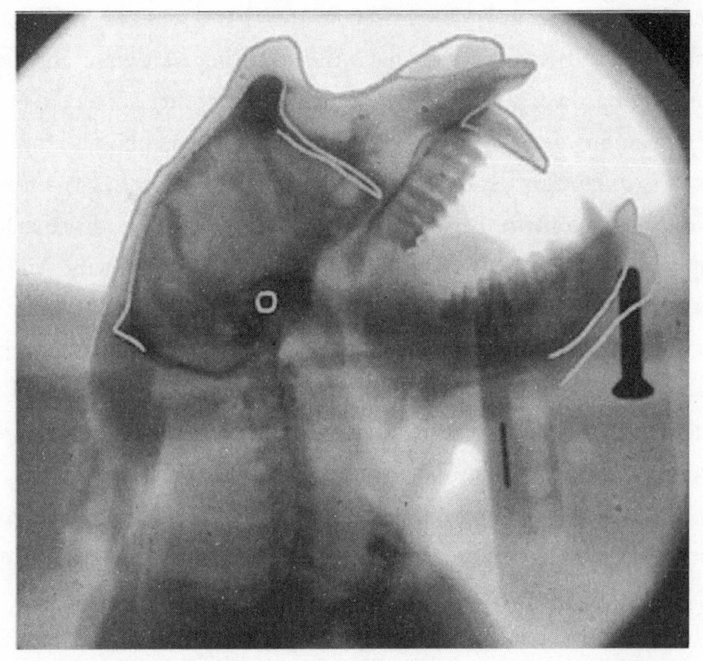

Emiliano hanging out in the scanner, showing off his vocal anatomy.

Tecumseh augmented this model with recordings of his wife speaking. He then automated his simulation of the monkey to see if its anatomy was capable of similar speech. And, in a high and somewhat whispery voice, thus spake Emiliano. Tecumseh used the test phrase "Will you marry me?" because the short sentence contains all the vowel sounds: I-O-U-A-E, which sounds pretty eerie whispered in a simulated monkey voice. The simulation showed that the monkey did, in fact, appear to have the anatomy to speak human words. This was a fascinating result because monkeys have never been successfully trained to do so.

So, how could this be explained?

We and many of our primate relatives, like chimps, all share the right vocal *hardware* to talk, Tecumseh said, but he believes humans are unique among the primates in having a brain connected to it. There is a neurological link missing in monkeys and chimpanzees, which makes it hard or even impossible for them to learn to control their voices. So, despite our evolutionary and anatomical closeness, we will never be able to have a spoken conversation with them. In Tecumseh's view, this is why we've failed to get research chimps to speak: They have the speaking hardware and perhaps the thinking hardware, but they are not wired together correctly, or at all. This theory, along with most things in the realm of speech and language, is contested (sometimes fiercely). Some researchers don't think the forebrain needs to directly control the sound-making muscles; others insist that we're missing ways in which other primates *do* control their voices; still others believe that we're measuring their voices wrong and that you don't use just your larynx and vocal folds (for example, I can make myself understood very easily by whispering, which doesn't involve vibrating my vocal folds). But whether or not our close relatives have what it takes to make sounds like us, so far we haven't found a way to get them to speak.

The vast majority of mammals and many of the birds that do vocalize are born instinctively knowing how to do so. These animals, from mice to chickens to squirrel monkeys, retain a limited, innate repertoire. They will all make the same sounds in similar situations, even if they are deaf. Humans, including those who are congenitally deaf, also produce some innate sounds, like laughter and cries, that we don't need to learn. But some animals also have the ability to hone and improve the use of their voice, often from observing or interacting with their fellow species from a young age. When an animal can actively learn how to change the sounds they make, it's called vocal learning. Baby bats, for instance, have a babbling phase where they make indistinct and simple sounds, their mother babbling back just like a human parent would to their baby: "*ababagogobaagogo.*" The baby mimics the parent, who encourages them as they train their vocal apparatus. Young zebra finches learn the song that they will sing their whole life with very little variation from older "tutor males" around them, practicing thousands of times a day. They can even learn their songs from watching videos of other males.

Some animals can even learn to mimic sounds made by other species with different vocal apparatus—like human words, even if they are only distantly related to us. An Australian duck called Ripper learned to say, "You bloody fool!" Koshik the Asian elephant learned how to stuff his trunk into his mouth (as humans put their fingers in their mouths to make whistles that their voices cannot make) and blow out Korean words such as "yes," "no," "sit," and "lie down." Hoover, a seal orphaned as a pup, learned to mimic his rescuer George Swallow's gruff New England accent to thousands of startled visitors in his home at the Boston aquarium. One of those was Roger Payne, who remembers how passing Bostonians would look around to see who had shouted, "Hey hey what are ya doin', what are ya doin'?" at them. According to Roger, "it was so convincingly human, so aggressively nuanced" that they'd totally miss the sneaky Hoover, who'd duck under the water as soon as he spoke the phrase.

How does this relate to whales? Well, the great masters of vocal learning, along with humans and songbirds, are cetaceans. This is intriguing when you realize that none of us are closely related; we have totally different ways of living and even different vocal apparatuses. Humans have a larynx. Songbirds have two-part structures called a syrinx, which enables some to actually sing two different songs at the same time, like a duet with themselves, while cetaceans have the extraordinary vocal tools I went into, literally, in the previous chapter. One famous cetacean mimic was the captive beluga whale called Noc. He could use his nasal tract and phonic lips to make sounds that were just like human conversations, despite human speech being many octaves deeper than the frequencies that belugas use to communicate. These imitations were so lifelike, the story goes, that a diver cleaning Noc's tank surfaced thinking he'd heard another human asking him to get out. Of course, being able to mimic human speech doesn't mean Noc the beluga had a clue what he was saying. So why did he do it? Noc had previously been a U.S. Navy beluga. His trainer there, Michelle Jeffries, believed that Noc "wanted to make a connection. I think that was part of the thing behind him mimicking speech."

Teaching other animals to speak like us had proven a dead end. But many of the captive primates tested for linguistic capabilities seemed to understand and respond to words spoken by humans, even if they could not themselves say them. As our closest relatives, they were still believed to be our greatest hope in understanding both the origins of language and interspecies communication. But unable to teach them our primary way of communicating—speech—scientists had to change tack.

Primates are highly communicative animals, great observers, and physical mimics, so instead of teaching them to speak, researchers used non-vocal languages to communicate. They taught them signs derived from American Sign Language (ASL), created ersatz symbolic languages based on systems of different tokens, or trained them to touch the screens of computer interfaces to select words and construct sentences. Chimpanzees, orangutans, and gorillas were all successfully taught these systems to crudely communicate with their trainers, and researchers looked to see if they could master elements of Hockett's "design features" that hadn't yet been discovered in primate communication systems.

Early results were promising. Chimpanzees who were taught ASL signs were then found to be teaching them to their own offspring. They signed to each other, to their keepers, and even to deaf children who visited them. In Des Moines, Iowa, Kanzi the bonobo learned four hundred lexigrams (abstract symbols representing words) and seemed to apply grammatical word order rules when using them. Kanzi and his fellow bonobos were studied in a thirteen-thousand-square-foot laboratory-cum-mansion, where they not only pressed lexigrams to communicate with their researchers, but operated microwaves, chose foods by pressing vending machine buttons, and selected DVDs to watch by pressing a computer screen.

Kanzi the bonobo "conversing" via lexigrams with Dr. Sue Savage-Rumbaugh.

The Central Washington University chimpanzee, Washoe, seemed to spontaneously combine signs to create new words. She appeared to show evidence of flexibility and innovation, word order preference, and an ability to talk about the past and objects and people not present. But while the chimps seemed to demonstrate special human language elements such as arbitrariness and semanticity, none of the systems the humans taught the chimps resulted in anything comparable to natural language as defined by Hockett's adherents. It was also all very one-way; the apes never used the techniques they'd been taught to ask questions. Perhaps they might have eventually, or perhaps they didn't care? Intriguingly, recent studies have shown that primates may have many other elements of natural language that we had overlooked with our focus on speaking. The chimpanzees of Budongo Forest, Uganda, for instance, were found to have at least fifty-eight unique gestures, used in a manner that follow some of the laws of linguistics and are "underpinned by the same mathematical principles" as human spoken language. Another team of researchers found that of the fifty-two body movement gestures one- and two-year-old humans use, such as head shaking and stomping, fifty are also found in chimpanzees.

What about other animals? Drawn by their extraordinary vocal imitation skills and repertoires, researchers have long been interested in parrots. For over thirty years, from 1976 to 2007 (when he died), the comparative psychologist Dr. Irene Pepperberg taught human language to her African grey parrot, Alex (his name was an acronym for "Avian Language Experiment"). She bought him as a one-year-old. He learned a vocabulary of a hundred human words, which he could say out loud. By the age of two, Irene told me, Alex could correctly answer multiple questions about the color, shape, or material of novel objects he was presented with. Color and shape, unlike dog or biscuit, are abstract concepts—they are ways we organize the world, and Alex seemingly understood them. This ability to switch categorization, Pepperberg explained to me, was a "demonstration of executive function levels of that of about five-year-old children." When given a mirror, Alex looked at his own reflection and said, "What color?" Having been told "gray" six times, he stopped asking. This was thought to have been the first time a nonhuman animal had asked a question.

As Pepperberg put it in 2016, "Not only had we achieved a kind of

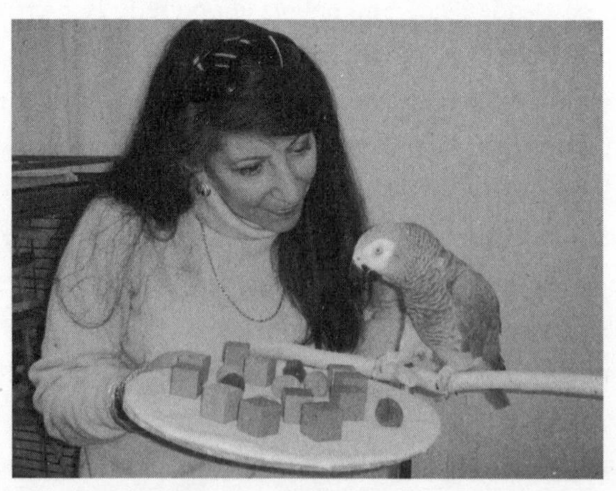

Irene Pepperberg and Alex, collaborating on an experiment.

'Dr. Dolittle' moment, but we felt we could be gaining insights into how language and complex cognition might have evolved in our ancestors."

It was during the same period of Irene's work with Alex the parrot, in the latter half of the twentieth century, that the debate over whether animals had language intensified. Of all these experiments with primates and birds, critics suggested some analyses and experiments were biased, and questioned the findings. Some reviewed footage and felt the subjects were reacting more to unconscious clues from the humans than responding to the tests, or that the humans were being too generous in their interpretations. Captive animal experiments fell out of vogue as the animal rights movement grew. One of the primate researchers, who had worked with the chimpanzee Nim Chimpsky, declared that he felt that Nim had just been making ASL signs in the hope of food, had not understood their meaning but had merely been aping him. Long-term research partnerships ended with the deaths of researchers or animals, and activists used the very discoveries about animal cognition that the experiments had enabled to campaign for an end to laboratory studies.

The work was time-consuming and frustrating. Some felt the results had been disappointing—the apes did not seem to have language as they defined it, or the capability to acquire it. Others felt complex communication *was* taking place and had been evidenced. Could this add up to something equal to human language? As Pepperberg found, "As more was learned, the bar kept being raised for the nonhumans." If animals could learn to use symbols for objects, that wasn't enough. They'd then have to be able to do so for verbs, too, and to combine them to make phrases, and then use this newly learned symbol system to demonstrate complex cognition by categorizing things they'd learned symbols for and showing how they related to one another, and so on. She described in frustration how other researchers "basically argued that language seemed to be defined as whatever it was that apes didn't have."

Science requires repetition. Alex the African grey parrot got answers to his questions right 80 percent of the time, but to get statistically significant results, he was repeatedly asked the same question, even after correctly identifying the answer. Apparently, he would grow tired of this and stop cooperating and then squawk, "Wanna go back," a term interpreted to mean that he wanted to be placed in his quiet roosting spot, where he could rest, safe from further interrogation.

The more I learned about this work, the more I felt pulled in different directions. The discoveries were so compelling and fascinating, the commitment of the experimenters to decades of painstaking work with their animal collaborators so impressive. But I could also understand why some were cautious of overinterpretation. There are hundreds of these experiments that you can watch and examine for yourself online. You can watch Koko the gorilla sign to her trainers about her pet kittens. When one tells her in English that a kitten has died, Koko uses the ASL signs "bad, sad, bad, frown, cry-frown, sad."

In these images, you can see Koko demonstrating twelve of her reported repertoire of 1,100 signs in American Sign Language.

I want to believe that what I am seeing is what I instinctively think: Namely, that Koko feels and thinks and communicates as I do, that she is there communing with her trainer. But am I just seeing what I want to see? A projection of what I would be thinking if I were in Koko's place? Are the videos really showing meaningful communication? Did Jack the baboon really understand that he was helping Jumper manipulate those signals? Did Old Tom the killer whale really think, *I'd better go wake those humans to hunt these whales*, or did he learn that slapping his tail fluke somehow produced dead whales to eat?

The more I read, the more I wondered whether discovering if animals had true humanlike language—however that might end up being defined—may have been getting in the way of something more profound. If animals think and feel in any way like us, and their communications open some window into their minds, might it not be worth trying to understand these and exploring how to interact with them better and more meaningfully, rather than attempting to test them for evidence of humanlike language?

I personally find it astonishing how *much* we have discovered with captive animals about attributes of natural language. This speaks to the commitment and ingenuity of the researchers. Teaching a zoo animal how to use invented human communication systems to pass tests with human inquisitors can shed light on that animal's innate cognitive abilities; you can control and replicate conditions, work with animals across their entire lives. But there are major drawbacks. These studies cannot give a window into how animals communicate in their real lives in the wild, where their communication systems evolved to work. Studies on individual animals raised by humans cannot discover the variety of communication that might exist between individuals and groups of those species. And if animal communication systems are taught and learned in their cultures, then how could you find them by taking animals out of those cultures? In searching for evidence of human natural language in animals, we had placed them into very constructed and unnatural situations. It therefore struck many biologists that there might be a more fruitful route to finding out whether nonhumans might have language, or something like it. Instead of teaching captive animals to use human codes, they would attempt to decode their wild communications.

For the Northern Arizona University Biology Professor Constantine "Con" Slobodchikoff, the need for proof that animals have complex communications is frustrating. In his book *Chasing Doctor Dolittle: Learning the Language of Animals*, he points out that scientists with pets they interact with every day "are literally surrounded with evidence that each dog or cat is a distinct individual, most of whom are plenty aware of themselves and their needs and spend much of their time trying to communicate those needs and desires to their owners." Yet this evidence doesn't count because it isn't recorded in a scientifically replicable manner.

If you have a hunch about what a signal from another species means, how do you find out if you're right? One test that scientists have developed involves sounding the alarm.

Alarm calls are very common vocalizations in the animal kingdom. Chances are you'll have heard lots of them. If you walk through a forest and listen for the birds, especially in breeding season, often they'll be repeating rapid sounds. You might think that what you're listening to is their song, the sound of that bird, but what you could well be listening to is that bird announcing *you*. Vervet monkeys also make different calls for different animals that scare them—leopards, snakes, and eagles all get different calls—and other listening monkeys act accordingly. When a vervet monkey hears another monkey make the call associated with leopards, it will run to the edge of a branch where leopards can't follow easily; if it hears the "snake call," it will stand up tall and scan for snakes; if it hears the "eagle call," it will flee to a less exposed position near the trunk of the tree. Scientists know this thanks to the success of what is called a playback experiment. They recorded the sounds of the different calls, then played them back to the monkeys over speakers and observed their reactions.

Most human communications—conversations, for example—are too complex to be able to unpick with playbacks, and most humans wouldn't respond well to having their conversations constantly interrupted and tested by experimenters blasting potential human warning signals. This may well be the case for lots of animal communications, too. But alarm calls are testable. For some biologists, alarm calls are Rosetta Stones for discovering whether there is semantic content within one animal's call that the listener can decode—that is, it's not just an emotional shout, but there is *meaning.* Much evidence has now been gathered from numerous playback experiments, confirming that alarm calls do contain information. I feel a little sorry for the countless animals that have been subjected to humans first scaring them with various frightening props—such as fake leopards and eagles—then playing lots of intense signals at them. Imagine if you were walking down the road and you heard someone shout "Flood!" or "Fire!" or revealed before you a giant stuffed bear. You'd probably freak out and climb a lamppost, look for a bucket of water, or cower behind a bush. Still, the discoveries from alarm call studies have been very intriguing.

Jungle fowl seem to have a variety of alarm calls in a vocabulary of at least twenty; caged chickens give different alarm calls for dangers from the land or the air, as do lemurs. And they are sensitive to being tricked: Vervet monkeys that were played an alarm call by the same individual lots of times (a "boy who cried wolf" situation) would ignore the call, but if a different voice was used, they would still flee. Impalas snort to alert others to the presence of a predator. When other impalas hear, they run away from the sound. Male impalas also grunt, which seems to be a sound used for competition with other males, who will rush toward the sound to take on their rival. Intriguingly, when they hear the grunt and the snort combined, male impalas run toward it much faster, the predator sound combining with the competitor sound to create a different, perhaps more urgent signal.

By breaking down the meaning-generating building blocks of nonhuman communication systems, the scientists who made these discoveries have believed that they challenged the long-held view that animal "language" can't do what human language can do. Some surprising evidence has come from work done with prairie dogs—social, ground-dwelling rodents that live in networks of tunnels. They're charming animals, greeting each other with kisses. In terms of vocabulary, prairie dogs have a bad rep. When they see a threat, they sit up on their hind legs and emit short, loud chirps, which to a normal person all sound the same. A comedy wildlife program in the UK even dubbed a video of an alarm-calling groundhog (a close relative of prairie dogs), making it seem like the animal was shouting, "Alan, Alan, Alan, Alan, Alan." The joke is that the simple sound is repeated so much as to be meaningless, so the animal seems foolish and unsophisticated. But this is a failing of human ears, not prairie dog "language."

Playback experiments and computer analysis found different calls for hawk, human, dog, and coyote. Different individual humans elicited unique calls. When the humans switched shirts, the prairie dogs kept the frequencies the same for size and shape but changed the one for color. If an experimenter fired a shotgun, or threw delicious seeds, the next time

A white-tailed prairie dog giving a "laughing bark" territorial call, so called because it sounds like a person laughing, Walden, Colorado.

they appeared the call the dogs made for them was different. These didn't seem to be built-in sounds made by instinct; instead, they seemed to be modifying their calls, even incorporating the color, size, shape, and incoming speed of the experimenter. Various domestic dogs were encouraged to walk through the prairie dog colony. The differently sized and shaped dogs elicited calls that seemed related to size and shape in humans, too, and the prairie dogs changed the speed of the call: If the dog was running around fast, the call would be faster. For Con, who pieced these findings together, it seems prairie dogs have nouns (human, dog), adjectives (big, blue), and verb/adverb modifiers (running fast, walking slowly).

The prairie dogs could also seemingly describe new objects. Con told me that he cut out silhouettes of coyotes, and the prairie dogs correctly made coyote alarm calls. When he then presented them with the outline of a skunk, or large triangles, squares, and ovals, he recorded new calls entirely. He writes they "seemed to reach into their store of descriptive labels, a vocabulary in their brain, and use that to put together a description of a completely novel thing that they had never seen before."

I've always wondered how bizarre playback experiments must be to some animals. Imagine if someone secretly started blasting recordings of other people in totally different situations to you—ten seconds of a market in Dakar, five minutes of some people from Yorkshire having sex, the shriek of an infuriated toddler—and then expected to gauge what the sounds meant from your response. But as with the alarm call work, some fantastic things have been discovered from playbacks. Do these prairie dogs think before they speak or are they acting out of sheer vocal instinct? As a biologist I know how important it is to tread carefully with any interpretation or comparison. But if Slobodchikoff is right, I find the implications of this to be staggering.

In 2019, Dr. Sabrina Engesser, from the Department of Comparative Linguistics at the University of Zurich, along with colleagues from the University of Exeter, published some results that hinted at how meaning is encoded in bird calls. They used playback experiments developed to see how human babies can discriminate between different speech sounds, and adapted them for birds called chestnut-crowned babblers, living wild in the Australian outback. She first showed that the birds could distinguish at least two individual sounds, which they called A and B sounds. Played by themselves, these were meaningless to the birds—but when rearranged into different strings of sounds—AB and BAB—the birds consistently reacted with different behaviors for each combination. When the meaningless sounds were put into different combinations, they meant something to the birds. Having meaningless sound units combine into meaningful "words" is what humans do, and until now it was thought that we are the only animal to do this. Sabrina's coauthor, Simon Townsend, wrote: "This is the first time that the meaning-generating building blocks of a non-human communication system have been experimentally identified." Sabrina's work with another bird, the pied babbler, seemed to find they even had some calls composed of sound units equivalent to "come to me" and others with those units changed equated to "come with me." What's more, Sabrina told me, these calls could even be combined with other threat calls into phrases, something like "come with me to this threat." Within a month, another researcher found similar results in the Japanese tit.

These insights into bird calls are so simple, even compared to how a toddler shapes their earliest words from the meaningless vowel sounds they first learn, but to me, they are electrifying. The findings from experiments like these break taboos on what we are prepared to think might exist in animal communications. We've looked in birds, monkeys, and prairie dogs. What more could we be missing in the vocalizations that are more complex, where the complexity has so far put us off probing further? "It's increasingly obvious just how much information is encoded in animal calls," says Holly Root-Gutteridge, a bioacoustician at the University of Lincoln. "There's now a preponderance of evidence."

Today, after some decades of study, no one knows for certain if any animal has language in the sense of natural language. It is perhaps wise to be skeptical of those with very strong convictions either way. There is a good chance that no animals do, just as there is a good chance that there may be no life outside Earth. My instinct, having observed human behavior in the wrangling over language, is that perhaps if we did find life on Mars, we would get into similar knots deciding if it should be fully qualified as life or not and miss the excitement of it being *something*.

Dr. Irene Pepperberg, looking back at the last fifty years of animal communication studies, lamented that arguments over language and methodology had distracted from the major discoveries of the field. Birds and apes had displayed such impressive capacities in the tests, mastering human-invented symbols and rules for using them, that to her it made sense that their own natural communication systems might be similarly sophisticated. She now prefers the term "two-way communication system" to the loaded term language, and I can see why. The general term most in use today is animal communication systems (ACS). The word "speak" is also troublesome. One biological definition for speech is "the preferred output modality for human language," and if you stick to this definition, no other animal can speak, because it is not a human. But what if you used a slightly looser definition of speech, say, "to vocally communicate your thoughts and feelings using linguistic units"? If a whale were able to vocally communicate its thoughts to me using linguistic units, I would definitely feel spoken to!

"How to have a two-way ACS with a cetacean using linguistic units" is, however, an awkward title, so I'll stick with "speak whale" as shorthand.

As I'd sat in conferences and eavesdropped, I'd also noticed a caveat. When scientists said, "Of course no other animals have language," some added the tantalizing bonus subclause, "with the possible exception of the cetaceans." For some, whether animals might have language was still an open question.

So, what would a whale need to do to be able to speak?

I formed a shopping list in my mind of helpful ingredients for an interspecies whale conversation. So far I'd learned that cetaceans had finely tuned ears, incredibly sophisticated voices, impressive and enigmatic brains, complex changing repertoires of songs, social lives based on their sophisticated vocalizations (encouraging for a human wanting to speak with a whale), and the ability to learn to change their vocalizations to imitate the sounds of other species. So far, so good.

What had we discovered about them that hinted at something more? I had asked Joy this question while she eased the brain of a minke whale from its skull. "Oh, you should totally meet Diana," she'd said. "She's been studying dolphin communication in her lab and in the wild for years, and she's just down the road in Manhattan, too!" And so, as simple as that, with an endorsement from Joy, I went to meet Diana Reiss.

7

Deep Minds: Cetacean Culture Club

Never trust a species that grins all the time. It's up to something.
—*Terry Pratchett,* Pyramids

The course of my life had been changed by a leaping whale, our brief encounter a gateway into tales of more profound and puzzling interspecies interactions. I had been awed by their speaking organs and felt their voices vibrate my own body; I had dove into the brambly history of animal language studies and emerged full of questions. What could we know about whales and dolphins from studying living ones? Did the scientists who knew them best think we might learn to speak whale?

Dr. Diana Reiss is a leading expert on cetacean behavior—someone with a window into their minds and communications. She had suggested we meet on her lunch break at CUNY, where she is a professor of cognitive and comparative psychology. I scanned the cavernous concrete-and-glass foyer, looking for someone who matched her photograph in the streams of students and backpacks. Diana saw me first. Before I knew it, she had introduced herself and we were off, weaving through the milling students and their reflections in the revolving doors at the entrance, and out into the city.

We headed down the street to a Jewish deli for lunch. Burly male waiters in white aprons whirled through the tiny space, leaning over us to pass plates to customers crammed into farther corners. I set my phone down to record our conversation but had no confidence it would be able to record a thing over the merrily shouting diners and clatter of plates and cutlery. Amid this hubbub, Diana would tell me not only of her groundbreaking dolphin research but also of film scripts she was working on, calls she had received from Leonard Nimoy (or Spock in *Star Trek*), collaborations with musicians and actors like Isabella Rossellini (who was one of her students), and the search for life in space. Although very careful and gentle in her movements, she seemed simultaneously to be in constant motion, like a dynamo. She was unlike most scientists I'd met, but then, as I was starting to realize, cetacean scientists were a strange bunch. Diana had first embarked on a career in the theater, but upon learning of recent breakthroughs—such as Roger Payne's work on whale song, and Ken Norris's study of dolphin echolocation—she threw over the stage for the lab.

Diana conducted some of the first-ever behavioral studies with cetaceans. She's spent decades observing and studying captive and wild dolphins, training them to use keyboards to communicate, showing them various objects and giving them tests to perform, watching and listening to them from birth into late adulthood. She has observed newborn dolphins hone their sound production, making their first, awkward whistle-squawks when they try to vocalize shortly after birth. By watching them, she saw how they had to learn to use their echolocation organs. For the first weeks of life, the babies were unable to "see" with sound and so relied on their other senses. When she presented them with new objects, they'd investigate them in ways more like a dog would—by coming to look at them closely and chewing on them.

Dolphins do some things that we long thought the preserve of humans alone. They plan ahead and use tools, choosing for themselves a sponge to protect their rostrums when pushing them into sand. A killer whale at MarineLand Ontario lured seagulls into his pool using bits of fish as bait. They play a lot: Wild dolphins pass seaweed and objects back and forth, seemingly teasing swimmers by keeping a piece of seaweed just out of their grasp. They bow-ride blue whales, surf and leap from breaking waves, pluck feathers from unsuspecting pelicans. Wild orcas circle and toy with human swimmers and frolic with kayakers. In captivity, cetaceans play with human objects, from frisbees to iPads. Two captive rough-toothed dolphins invented a game in which they would take turns towing each other with a hula hoop around their pool. Perhaps the most complex example is unique to cetaceans—bubble play. Diana's dolphins and others will exhale perfect bubble rings, attending their creations like potters with their wheels, topping up an uneven ring with delicate exhalation. Some then propel them around sideways underwater with their tails, crafting a series of spiraling rings, which they'll swim through.

A dolphin creates and then inspects a double bubble ring.

A perplexing dolphin cognitive ability is pointing. Most animals do not seem to comprehend pointing, apart from dogs and, surprisingly, bottlenose dolphins. Dolphins were able to understand pointing commands when their human trainers used fingers or arms pointing in a different direction than the human was facing, and even a sequence of points at different objects with a command sequence that could only be interpreted correctly in one way ("Take *this* ball to *that* basket"). No other species can do this. In the words of biologist Justin Gregg: "Why dolphins—an animal that has no arms, hands, fingers, or any other appendage that could produce something resembling the human pointing gesture—should have this ability is still a mystery." Dolphins can point themselves by holding their whole body motionless and aimed in a particular direction. Captive ones point things out to their trainers, while wild dolphins have been observed pointing out dead dolphins to one another.

Diana works both with wild dolphins in Belize and Bimini and captive dolphins at the New York Aquarium and the National Aquarium in Baltimore. When working with captive dolphins, Diana stands facing them at their level in the tank through a giant glass wall. Video cameras and sound recording devices are rigged to analyze what the dolphins do, what sounds they make, and in what order. To my ears, it sounded eerily like the science fiction film *Arrival*, where a linguist is assigned to try to decode the communications of two aliens that have arrived on Earth in a spaceship, but is only able to signal through a transparent wall. Diana told me that while dolphins had a different kind of intelligence, and bodies different from our own, it had surprised her to discover they were very similar to us in some ways. I asked what she meant. She said she found that in the kinds of emotional reactions her dolphins expressed, how they responded to her tests with mirrors and special underwater keyboards, they seemed in some ways human. Young dolphins learning how to communicate appeared somewhat similar to children acquiring early elements of language, "though I don't call it language with the keyboard," she added carefully. "Yet, there are other things that are... You're just stumped."

As mentioned earlier, dolphins are vocal learners and excellent mimics. Each dolphin has a completely distinctive "signature whistle," a learned vocal label that is thought to function somewhat like a name, a way for dolphins to refer to and address each other when they meet. They often copy each other's whistles and can remember their friends' signature whistles for more than two decades (both in the lab and in the wild). Dolphins have been heard doing impressions of humpback whale song; and Guiana and bottlenose dolphins, for unknown reasons, mimic each other's sounds when they fight. This imitation seems a widespread trait in the toothed whales: Killer whales have been observed imitating other whales and even other noncetacean species. Some orcas have learned to make barking sounds like sea lions. So Diana was not especially surprised to find her dolphins mimicked computer-generated sounds that were like nothing else they had ever heard.

One of Diana's captive dolphin experiments involved an interactive underwater keyboard she designed, with white visual symbols on a black keypad. If the dolphin pressed one of the symbols, an underwater speaker would play a novel electronic whistle, different from the whistles the dolphins themselves produced, and something specific would happen. For example, the animal would hit a symbol for "ring," and they'd hear a specific whistle and then receive a ring; or they would hit a different symbol for "ball," hear a different whistle, and get a ball, and so on. Unsurprisingly, for one of nature's great mimics, the dolphins quickly started imitating the sounds. But then, Diana said, one day the dolphins were in the pool doing a different experiment, and the keyboard was not present. The dolphins were playing with a ball and making the ball sound, and playing with a ring and making the ring sound. She described how this was similar to "what children do when they play with toys." No one was rewarding them with a fish for their participation. They'd incorporated her symbolic computer sounds into their own communications.

Things got even more interesting from there. The dolphins started pressing both the ring and the ball keys simultaneously, wanting to play with both toys together. At around this time, they began making a new kind of whistle, which the scientists didn't recognize. It was only by examining the waveform on the computer screen—a graphic representation of the sound—that they realized the new whistle looked like the waveforms for the ring and ball sounds combined. "Ring-ball." The dolphins had never heard the sounds joined together (the computer played them with gaps). This was a big deal for Diana. She was seeing dolphins take her signals, learn what they meant, learn to say them, and then combine them into a new signal, by themselves, unprompted. I asked Diana how she felt at that moment. "I loved it!" she said. "But I was very careful."

While all scientists should be rigorous, I was beginning to get the feeling that many cetacean scientists are extra careful. This owed something to the complex legacy of a legendary New Age figure, Dr. John Lilly. A controversial and fascinating man, Lilly began his career as a neurologist. His vocational interests whorled out from contributions to conventional science, physiology, and psychoanalysis to cognitive experimentation with LSD and ketamine, the invention and use of sensory deprivation chambers, and the study of dolphins and their "language." Friends with Beat poet Allen Ginsberg and Timothy Leary, a prominent advocate of psychedelics, Lilly made midcentury discoveries about dolphin physiology and anatomy that brought the cetaceans—long neglected by researchers until then—much wider scientific interest by the 1960s. But, in parallel with Lilly's experimenting with LSD, his work gradually turned further from relatively classical scientific research into theories of dolphin higher consciousness and telepathy, as well as theories of how humans would eventually be replaced by thinking machines, which today seems perhaps less far-fetched. At one stage, he set up a half-submerged house-cum-laboratory in Florida where a human and dolphin could live together, to study dolphin language and interspecies communication. What didn't help the program was that Lilly injected LSD into one of the dolphins at the laboratory, or that one of his assistants masturbated their research dolphin, Peter, when he was sexually distracted. Funding dried up and the program was halted nine months in, although researchers there felt Peter was making progress in mimicking human speech and sounds.

In this picture, researcher Margaret Howe and Peter the dolphin in their semi-submerged lab. When they were separated at the end of the experiment and Peter was moved out of the lab, he reportedly killed himself.

For some, Lilly had progressed "from a scientist with a white coat to a full-blown hippy." As Dr. Diana Reiss describes it, "Some of that work was very controversial and sort of pseudoscience-y after a while, and highly speculative." This discredited his later dolphin research, as well as his earlier, scientifically published work in the eyes of many, and threw up a cloud of skepticism that dolphin researchers would have to negotiate for decades to come. Many are still at pains to distance themselves from efforts to "understand dolphinese," for fear of being seen as unserious. Justin Gregg notes that these days, in part thanks to Lilly, "There are probably more weird ideas about dolphins swimming in cyberspace than there are dolphins swimming in the ocean." And yet Lilly was a trailblazer, without whom scientists such as Diana Reiss, despite treading carefully around his later work, would not have been inspired to join the fray.

This history was why, after her dolphins seemed to have created a new whistle for "ring-ball," Diana scrutinized hours of recordings to make sure. It was no coincidence. The dolphins conjoined the signals in twenty-eight sessions, and in all of them they were playing with the two toys together. "It's what we call behavioral concordance," she explained. The sounds had matched the behavior. We also conjoin words for things when combining them. Foot-ball. Shot-put. Irene Pepperberg had observed something similar with Alex, her legendary African grey parrot. Alex liked corn and knew the word for it, and therefore would say "corn" to request it. Irene had run out of yellow corn, so one day she gave him Indian corn, which was harder. When Alex bit into it he said "rock-corn," combining the word "rock," which he also knew, to signify this new, horrible kind of corn. The chimp Washoe, trained to do sign language, was out on the lake with Roger Fouts, the scientist who studied her. They saw a swan, a bird Washoe had never seen before, and the chimp signed "water-bird." If these reports are accurate, these were innovations in communication—animals not just repeating what they'd been taught, but using words and reshaping them for new expressive functions.

There is, of course, always the danger that the scientists had accidentally trained their animals to do this *unconsciously* through a process called reinforcement, which those working in animal communication experiments must be very wary of. Who is to say that Alex and Washoe hadn't come up with their conjoined words by chance, and, seeing the excitement of their watching humans, learned to make these sounds again, with no understanding of what they meant? In the case of Diana's dolphins, she explained to me that the scientists couldn't have been a factor because they couldn't react. They hadn't noticed at the time, only realizing the dolphins had created combination sounds seconds later, when they were analyzing the recordings. What could we infer from this? Could these dolphins' readiness to combine the signals she'd taught them indicate that their own communication signals might be combinatorial in nature? "We're still in the infancy of this," she said. But what is definitely true is that over the last few decades, through experiments with trained dolphins, we've glimpsed cognitive abilities we didn't think any animals other than humans had. A mother dolphin and her two-year-old calf placed into separate tanks and given an underwater telephone to communicate could happily chatter away, vocalizing back and forth in turn.

One of the most prolific scientists in the field of dolphin science was a man named Dr. Louis Herman, who worked in Hawai'i. Here, he trained two dolphins, Akeakamai (Ake) and Phoenix to use complex communication systems. One dolphin's system was based on various gestures the trainer used, whereas the other's was made up of various sounds. Ake became so good at using this system that his trainer could ask him to follow commands made of not just one sign but whole sequences of them corresponding to different objects, positions, directions, relations, and agents in and around his pool. The sequences of up to *five* symbols, including actions like "fetch," objects like "basket," and modifiers like "left" or "right," only made sense once they had been fully transmitted, so the dolphin had to wait to hear or watch the full sequence before it could understand the relationship between the multiple objects in the "sentence," and then make sense of them by carrying out the command correctly. For instance, "Right water, left basket, fetch" meant taking the basket on the left to the stream of water on the right. They were also taught a "question" symbol, indicating that the dolphin needed to answer using a yes or no symbol. They could press paddles to confirm to the trainers that the objects they were asked about were there and could even remember to perform tasks with objects that weren't there until after they were given their instructions. "The dolphins could respond correctly to novel sentences that used familiar words given in a never-before-seen sequence," writes Justin Gregg. When they were given intentionally wrong symbol sequences, they either didn't respond or "extracted a meaningful phrase by ignoring certain elements while retaining the semantic relationship for word order."

As I'd learned, being able to use syntax and infer meaning from symbols are elements of our own human natural language. In other tests, dolphins indicated comprehension of further concepts; they can classify objects based on their shape, number, and relative size. They classified a concept of "humans." It's hard to explain these achievements without dolphins having a brain that can make mental representations of things. And if they could do so in a pool, with human-wrought symbols and odd and repetitive tasks, then why not in the wild with the objects, relationships, and environments of their evolution, where their survival could depend on it?

A diver and a bottlenose dolphin at the EPCOT Center underwater keyboard in Orlando, Florida. This early interface could be operated by human or cetacean. You can see the diver breaking the infrared beam to generate a word in English, with the dolphin paying close attention.

It is partly these glimmers of insight from cetaceans given communication tasks that cause some biologists to pause when ruling out language in nonhumans. And not just this, Diana explained, but also other discoveries about their cognitive capacities. The blobs of creamy tissue I'd seen a few blocks away—what sort of inner world were they capable of constructing? What kind of mind would you encounter if you could speak whale? Diana and her colleagues have been investigating this, too. She began with a mirror.

Before I sat down to write this sentence, I poured myself a glass of water in the bathroom and looked up at the mirror above the sink. I noticed that I had a black smudge on my forehead from some garden rummaging. I reached up and wiped it off. I had just passed what is known in biology as the Mark, or Mirror Self Recognition (MSR), test. It is thought to indicate that I can conceive of my own identity, that I recognize that what I see in the mirror is a reflection of me, and also that I know what "me" is—an indication of self-awareness. Until recently, having a concept of self was thought to be another one of those things unique to the human species.

If you present another animal with a mirror, it may act in many different ways. It is a visual test. Some animals have such different senses that they cannot behold themselves in a piece of reflecting glass. A worm will not respond because it does not have eyes. And even if the creature in question has eyes, they might not see as well as us with our depth- and color-perceiving ones. Some animals presented with mirrors will register their reflection, but react as if it's another of their own species. Siamese fighting fish will attack the mirror, supposing it to be a threat. This also seems to be what happens when long-tailed tits in the UK knock on our windows: They are not summoning our attention, but pecking at reflections of rivals. Or so we guess. They might just find mirrors confusing, not lack self-awareness.

Other animals, however, seem to understand that it is themselves they are seeing. They will tilt their bodies, move back and forth with eyes trained at their self-image in ways they wouldn't act toward their peers. These behaviors are classified as "self-directed." Few mirrors exist in nature, and few evolutionary forces will have primed animals to react appropriately to a reflected image of themselves. Figuring out what that image is, is considered diagnostic of what your brain can do. Thus, to see yourself, to be "self-aware," is thought by some to be one of the benchmarks of consciousness. To test whether some animals really do see "themselves" in the mirror, scientists devised a clever further test. When the animal was unaware, they would mark them with something, like a red smudge of chalk or spot of dye, often on their head. Like my smudge of soil. If the animal looked into the mirror and beheld its reflection, and on this reflected animal it noticed and investigated this unusual mark, it would be evidence that they knew who they were looking at: themselves.

Many scientists felt that only apes would pass the MSR test. But in 1987, Diana took a mirror and showed it to some bottlenose dolphins. Dolphins have no necks to write home about and their eyes are on the sides of their heads, so they can't see most of their own bodies. The two young male dolphins showed interest in the mirror and appeared to use it to view themselves. Later they performed "sequential intromission attempts" in front of the mirror—meaning they had sex and watched themselves.

This was not the experimental setup, but it is a really nice picture of Diana and her dolphin.

In 2001, Diana and her colleagues took the experiment further. They drew marks on the bodies of a different pair of dolphins with a pen—above their eyes, behind their pectoral fins, near their belly button, and so on—and watched to see what would happen. They found that the dolphins would head to the mirrors after they'd been marked and twist and flip themselves in the water, appearing to closely examine the reflection of whichever body part was marked. When Diana's team drew on the tongue of one male dolphin, he went immediately to the mirror and opened and closed his mouth repeatedly. To top things off they demonstrated other "self-directed behaviors": blowing bubbles, turning upside down, and wiggling their tongues, all while hovering intently in front of the mirror. These behaviors feel much more relatable to my human sentience than noticing smudges!

The age at which dolphins are able to demonstrate such mirror sentience is impressive. Bottlenose dolphins aren't too different from humans in some respects, with females living to around sixty and reaching sexual maturity sometime before the age of fourteen. In another study in 2018, Diana found that a young dolphin called Bayley was able to mirror self-recognize at seven months! This is younger than human children, who generally start being able to recognize themselves in the mirror at about twelve months, and chimps at around two to three years.

Her findings caused a surge in interest in both cetaceans and in the MSR test, and many other species previously thought unlikely to pass were tested. So far, our close relatives, the chimps, orangutans, and bonobos, all pass, but slightly more distant cousins such as Barbary macaques and other monkeys fail. Elephants pass. Dogs fail. Cats fail. Sea lions, pandas, gibbons, African grey parrots, crows, jackdaws, and great tits all so far have failed. Diana's hypothesis that dolphins are self-aware was supported when their larger relatives, the killer whales and false killer whales, were given mirrors, and they seemed to react to their reflections as I would if I saw mine for the first time. They bobbed their heads and watched their reflections moving, checking out their tongues and the other bits of their bodies they couldn't normally see.

The MSR results with dolphins have bolstered arguments that dolphins are self-aware, even as what being self-aware might mean to the variety of life on Earth has become harder to define. And MSR-based research is intriguing. It leads into a thorny field of biology where philosophers roam, studying consciousness itself. Trying to figure out how to explain why we have "the feeling of being inside your head, looking out" is so difficult that it has been referred to as the "Hard Problem of Consciousness." Suffice it to say that "consciousness" is actually so hard to define and culturally argued over that, like "language," it's easier to avoid in favor of more straightforward but unromantic terms like "cognitive functioning." And for researchers like Diana, mirror studies are simple devices that allow us to get a reflection of animals' cognitive functioning ability.

There are now hundreds of papers on behavioral experiments that shed light into the dark recesses of cetacean minds. There's evidence that dolphins can conceive of having a body, that bottlenose dolphins and killer whales can choose what activity they'd like to do next (a facet of free will), and even, when prompted, invent a new task to perform—which is hard to explain if you think they are reflex-driven biological machines. But what does this all mean for the intelligence of cetaceans in general? It's worth remembering for a moment that almost all these discoveries have come from a handful of humans testing a handful of dolphins. Mostly bottlenose dolphins, mostly in captivity. A few individuals, in a strange environment, with a few tests, to represent the capacities of millions of different individuals of ninety-odd cetacean species.

Diana Reiss and Louis Herman might have somehow had genius-level individual dolphins, and perhaps bottlenose are the species closest to us in the cognitive and communication abilities we so admire, with ideal tests for picking up their talents. Yet it's unlikely that they are. What is more likely is that nosing through tropical rivers, leaping from temperate seas, and passing silently under icecaps are cetacean intelligences as varied as the cetacean bodies they inhabit. What we glimpse through Diana's work and that of others like Herman is a fraction of what some of the brains of some individuals of some species can do. I also had to remind myself that with any discovery in one species of this extraordinarily diverse bunch of animals, we should be wary of extending the assertion "some cetaceans can" to "cetaceans can." What a captive bottlenose can demonstrate does not translate to what a blue whale or Cuvier's beaked whale or river dolphin is or isn't cognitively capable of. It is just a beginning, but it is an exciting one.

Speaking to Diana, I was struck not only by the challenging nature of her work but also how vital she felt it is. Her hard-fought insights into the minds of cetaceans reinforced her conviction that there was somebody home inside the bodies of whales and dolphins, and that other people should know about this. "For me," she said, "this is translational science," a means of scientific discovery that leads from what we know about an animal directly into the ethics of how we treat them.

Our lunch was drawing to a close. Diners we'd initially been sitting next to had left and been replaced by others, and our waiter hovered, eyeing our empty plates. Diana's phone rang; she would have to leave shortly. She told me she wanted to share one last story, the one that had had the greatest effect on her life. It was not about a dolphin, and it wasn't even from her laboratory. It was the story of how she had once communicated with a wild and lost whale.

In 1985, Diana was conducting dolphin research and teaching at San Francisco State University when she heard that a humpback whale had entered San Francisco Bay. A busy shipping channel, it was a terrible environment for a whale to find itself in. Swimming up the Sacramento, eighty miles inland and upriver, the oceanic whale was in a freshwater habitat, and scientists were worried he wouldn't be able to find anything to eat, and that his buoyancy and skin would suffer. The whale, named Humphrey by the media, was soon a global sensation, with helicopters tracking him from the sky and news reports enthralling the nation. But he couldn't seem to find his way back out to sea and his days seemed numbered.

Part of Diana's work involved advising the nearby Marine Mammal Center, and she joined the rescue effort. The rescue team tried everything to get him back to sea, from banging metal pipes in the water, as Japanese dolphin fishermen do to herd them, and even playing the sounds of orcas to scare him toward the ocean, but both had little effect. One of the government officials involved tossed a seal bomb—a sort of underwater sound grenade used to scare sea lions away from fishermen's nets—and poor Humphrey beached himself and had to be rescued. Diana remembers looking into his eye, pouring water on him, trying to keep him calm. Once he was back in the water, the team decided to change tack; they would stop trying to herd him in the right direction and instead try and entice him, by playing recordings of the sounds other humpback whales made to each other while feeding in Alaska. Diana and the team headed out with an underwater speaker in a Zodiac semi-inflatable boat. Humphrey was nowhere to be seen, even from above. But as soon as she turned the tape on and played it, he appeared out of nowhere and followed her boat for eight hours. He seemed drawn to the sounds of the other feeding whales.

Only the night before, Diana had gone to her lab and watched the dolphins in the pool, and she'd realized that when the dolphins were together, they tended to be quiet, communicating when they separated. She decided to try the same tactic with Humphrey. "When he was near, I shut off the sounds. When he started wandering away, I turned them on again—it was like calling your dog. He came right to the boat. It was unbelievable. It was the first successful playback experiment we have ever had."

Over the following day, Diana's team continued their playbacks, luring Humphrey farther and farther out of the bay, and finally out under the Golden Gate Bridge. Once they passed the bridge, they lost him completely. Diana told me how she instructed the small fleet of a dozen boats to turn off their engines, and they waited and watched for him in silence. Suddenly Humphrey appeared right next to her. He pressed his belly to the side of her boat and looked up at her and the other members of the whale rescue team. "That was the most remarkable thing I have ever seen," she said. The humans hung off the side of the boat, teary-eyed, until Humphrey turned and headed off in the direction of his peers. It had been almost forty years. Diana seemed totally transported by the memory of saving Humphrey. "I think at that moment, there was some real communication that went on. It was one of those stellar moments in your life." She paused. "And this *says* something about these whales."

If the playback experiment worked—and it seemed to—and if the whale calls Diana had broadcast underwater had held some sort of whale meaning that had been understood and responded to by Humphrey, then there is an argument to be made that this is the first time in history that someone had spoken whale. Albeit via a machine, with a recording of another whale and with no idea of what they were saying. Equally, Humphrey could just have heard the sounds whales make when they eat fish and, starving and alone, decided that he wanted to be near those whales and those sounds. But it seemed to me there was no doubt in Diana's mind that she was dealing with an animal capable of complex communication, perhaps akin to language. Why else would she have felt so moved by her connection to Humphrey?

Our time was up, and Diana left me in my debris of notepapers and sandwich crumbs to rush back to her students. I thought how the singular moment of her research life that she'd wanted to tell me about was not in a lab, but out at sea with a lost whale and the moment he had come to look at her. It was as if the power of this experience—out of control, on the terms of a wild animal—was as valuable to her as the accumulation of laboratory evidence.

Until the last few decades, the question of whether we might communicate with animals would have been met with derision. For centuries the limitations had been with our culture: We didn't care about other animals, like whales and dolphins, and we didn't think it was worth finding out too much more about their inner worlds. But now we do care, and we are fascinated. We now know that these animals communicate, not only with their own kin but also with other species. We know they have bodies built for communicating, and impressive brains. In laboratories, these bodies and brains seem to be able to learn ways of communicating that hint at impressive cognition, a certain grasp of some of the elements of our speech and of our conceptual universe. The whales and dolphins are out there, communicating, perhaps even conscious. If any other animal was going to be having conversations, they seemed a good bet.

So how could we find out for sure? Diana had laid out the scale of the problem to me:

"We have no clue how their vocalizations are organized and how they function." She had reached over and drawn the contour of a dolphin whistle on my notepad. "Is this a sentence or is this a word?" she had asked. What was even the beginning and the end?

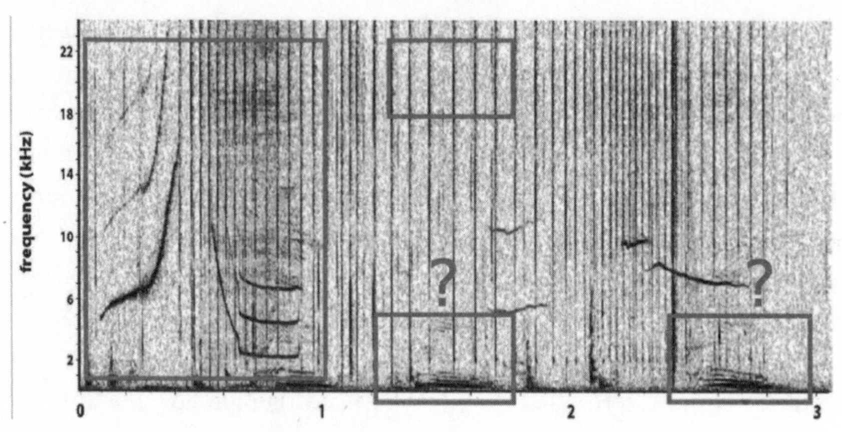

Spectrogram of dolphin vocalizations recorded by a student of Diana's.

What were the dolphins' signals among the hundreds of thousands of hours of recordings, and what was noise? Imagine if you wore a microphone all day. You'd make a range of different noises. Some of them would be your communication noises, but there would also be grunts, stomach rumbles, sniffs, and sighs. How could a dolphin scientist studying the recordings from your microphone know to distinguish the meaningful noises you make from the humming, unfinished sentences, tuts, and belches? It's one thing to try to break a code, but first you have to sift it out of the noise. Even if you knew, it would take more time than you had in your life to go through and highlight the sections of recording to investigate. You would need thousands of lives. And we had only recorded a fraction of the sounds an individual wild dolphin or whale made in a day and matched this to their behavior, let alone a group of them.

When I thought of the challenges of trying to get inside another mind, one that lives underwater and sees in sound, I could understand why we could not speak whale yet, and why we'd tried to teach them to speak in ways we could already understand—in picture symbols, in human sounds. Having spoken to so many people who knew these animals, who studied them and had spent their lives trying to be around them, I'd come to wonder whether the main hurdle to understanding them was us: our limited senses, bodies, life spans, and minds.

They lived in the sea, where our senses do not work effectively and where we cannot breathe. We could only venture out on boats and only then catch glimpses of their lives. How could we hope to understand them if, for most of the time, we couldn't perceive or record them? And if we could somehow glean enough relevant information from their wild lives, how, with no Rosetta Stone, could we have a chance of decoding their communications?

I had been skeptical at first that we would ever have a chance of understanding cetaceans. As Joy had said to me at the outset of my journey, "You can't just ask a whale." And this is true, and always has been. But while Diana and her colleagues pushed the boundaries of what humanlike capacities we could discover in dolphins, other scientists had been developing ways to spy on these cetacean minds in their own worlds, on their own terms. At the beginning of the twenty-first century, they began to bug the sea. They created mechanical minds and bodies that were superhuman, setting biologists free of their previous limits.

If everything so far has been about the past, what follows is about the future.

8

The Sea Has Ears

Nānā ka maka, hoʻolohe ka pepeiao, paʻa ka waha.
Observe with the eyes; listen with the ears; shut the mouth.
Thus one learns.
—*Traditional Hawaiian proverb*

In 1967, the poet Richard Brautigan imagined a future where machines liberated us from our toils, allowing us to return to our "mammal brothers and sisters" all the while "watched over by machines of loving grace." As our conception of what animals are has changed over time—from being mere biological machines to sentient, even intelligent beings—machines have been transforming our capacity to *perceive* animals. This shift has accelerated rapidly in the last few decades with technologies that give even the most urban, nature-starved human the opportunity to observe the lives of other animals as never before. Brautigan's prescient vision has not quite entirely come to pass in the way he may have imagined, but the machines are certainly here, and they are watching. And listening.

The island of Hawaiʻi, or the Big Island, is the largest in the Hawaiian archipelago, a chain of 137 islands, atolls, islets, and seamounts that scatter northwest to southeast across 1,500 miles in the middle of the Pacific Ocean. It is a young land, a giant volcanic mountain formed from magma spilling up from the guts of the Earth, spurting through the planet's crust

A humpback whale spy hops, looking at its surroundings above the waterline.

into the deep sea and accumulating above the surface into a ninety-three-mile-wide mass of extinct, dormant, and active volcanoes. It is alive and growing. I watched lava pouring from its cliffs into the ocean at night, the hot red rock and white steam illuminating the island's walls, and drove through suburbs where mansions and roads were rudely truncated by great piles of hot, jagged rock, the houses that had stood there months before indicated now only by their postboxes. New roads built over the fresh lava steamed constantly, great fissures vented stinking acidic gases, slowly dissolving the metal gates and tin roofs of neighboring surviving houses. It seemed otherworldly.

The highest mountain is the sacred peak of Mauna Kea. According to Hawaiian creation tradition, when the islands were created by the Earth goddess Papa (short for Papahānaumoku) and Sky Father Wākea, they began with the Big Island of Hawai'i. The mountain was their firstborn and its summit—the piko, or navel—its center, beginning, and end. If you measure it from the deep-sea floor where the mountain begins, to its peak at 13,803 feet, it is the tallest mountain on Earth, bigger than Everest. You can swim in the warm ocean that laps its sides, and then drive up to the top, euphoric and giddy with the altitude and, once there, crunch across its icy crown. You can watch the tropical sunset, your view framed by the golden orbs of some of the most powerful telescopes on Earth, whose scientific stations pepper the island's mountaintops, their covers clunking open in the gloaming. The air is thin, and there is little light pollution. From these stations the accumulating carbon dioxide in our atmosphere has been tracked since 1958, and out of these telescopic lenses we have peered into deep space at the atmospheres of distant planets. Across the slopes below are deserts and lava plains, lush jungles and snow, black sand beaches and swamps. The weirdness and diversity of its geology is matched by a hodgepodge of human settlement. The beachfront properties of tech tycoons, scientific research outposts, and Native Hawaiian protest camps against their expansion, coffee plantations, resorts, clitoral meditation retreats, cattle ranches, manicured and deathly quiet golf courses, U.S. military installations.

Hawai'i might look like paradise, but it is home to one of the most disturbed ecosystems on the planet. The coming of the first Polynesian settlers 1,500 years ago followed by the European waves of the last few centuries brought an ecological whirlwind, laying waste to the unique plants and animals that had evolved there. Animals were hunted, forests were removed, and in their place, new animals and plants were intentionally introduced or arrived by accident—stowaways colonizing the island chain. So many species were lost that it has been called the Extinction Capital of the World. In 2019, the year I visited, a tiny survivor of these previous extinctions, called George, had just died. He was a Hawaiian tree snail, the last of his species, *Achatinella apexfulva*. In the nineteenth century, records claim 10,000 snail shells could be collected in a day in a Hawaiian forest, with more than 750 species counted. Less than one third of them remain alive today. In Hawaiian legend, snails were revered and thought to sing beautifully, but no singing species survive. Other extinct beasts include the Kona giant looper moth, the silt-owls, the Laysan honeycreeper, and a small bat we know of only from the discovery of its tiny bones in a lava tube. While I was writing this book, another 11 species of birds were declared extinct. The forests of Kauai will no longer ring to the haunting call of the ʻoʻo; no one will glimpse the brilliant colors of the Maui akepa, or Molokai creeper.

Most species died before the waves of Europeans with their classifying fetish had discovered and bestowed Latin names on them—extinct animals defined in an extinct language. Bird species alone crashed from 140 to less than seventy, with the survivors trapped in tiny fragments of forest. On the Big Island many of these survivors now lived high up above the golf courses and the hotels, above even the cattle pastures, green deserts of boggy grass where once were rich and complex forests. The trees that weren't cut down grew old and their seedlings and saplings were snaffled by the relentless mouths of cows and sheep, so no new ones grew to replace the old. Now, only the odd, ancient stand of pre-European forest remains. But even in their refuges in the few remaining groves, the birds were not safe. Invasive mosquitoes brought with them avian malaria, and as climate change warmed the mountains each year, the mosquitoes' range crept higher; as they fed on the surviving birds, they passed on their deadly parasites. There is now nowhere left for the birds to go, squeezed between a plague below and thin air above.

It may seem surprising, but this island chain, the birthplace of surfing, the last state to join the American union, is also a test bed for machine animal spies and listening devices. These tools have been pioneered by scientists studying rare animals, hard-to-find animals, animals for whom the stakes could not be higher and who must be undisturbed, and animals on the brink of extinction, tiny birds and colossal whales. Visiting Hawaiʻi was an opportunity for me to meet some extraordinary people whose work developing technologies to save endangered species could also be instrumental in cracking the codes of animal communications. At the University of Hilo on the east side of the Big Island there is a laboratory called the Listening

Observatory for Hawaiian Ecosystems (LOHE), the acronym doubling as a Hawaiian word that means "to perceive with the ear." It is a bioacoustics lab that specializes in deploying listening devices honed to detect animal voices, in places where humans cannot go or spend significant time.

This is their logo, featuring a humpback whale and an ʻiʻiwi bird beneath a snow-capped volcano, united by an acoustic waveform. In a crowded field, it's my all-time favorite laboratory logo.

The LOHE logo.

It was pouring rain in Hilo when I met Professor Patrick "Pat" Hart. He had tousled salt-and-pepper hair with laughter lines by his eyes, which leapt into action alongside the big grin he greeted me with. We ducked through the deluge into a local supermarket to buy snacks for what would be a long day ahead. "Don't eat the salad," he warned me. "There's a rat lungworm outbreak and you can get it if you eat a slug." I didn't need any further encouragement. Pat was joined by a fresh-faced male grad student from Baltimore named Andre, and we were off to visit their remote field station in the Hakalau Forest National Wildlife Refuge, where some of the world's rarest birds lived. Andre told me he was a Vietnamese American and that he'd defied his parents' wishes and switched from studying marketing to animal behavior. It was Andre's first trip up the mountain, and he drank in the landscape as we drove to what would be his research base for the next few years. The reserve sits high on the flanks of one of the two main volcanoes, and to reach it you ascend from tropical coast, with jungly trees, through grassy plains and across rocky volcanic slopes, before slipping back into more boggy grassy areas, where the mountainside seemed to trap the misty air. The landscapes were green but bare and treeless as a Mongolian steppe. After a couple of hours bumping down a rocky path, we began to see, in patches here and there, huge, dark, and wrinkled trees. I realized that, until this, we hadn't seen any foliage really for a long time. These 'ōhi'a trees had been here since before the Europeans, and were perhaps four hundred years old.

As we drove, Pat told us about the wonders of Hawaiian birdsong; how much more complex their calls are than our human voices, which he dismissed, jokingly, as "just low-frequency mumbling, really." He told me of how different bird species will sing at different pitches and times—in different acoustic channels—so they didn't speak on top of each other. We'd long assumed birdsong was simple, but the more Pat looked, the more complex, individual, and changeable it appeared to him.

After a good couple of hours on the road we rounded a bend and found ourselves driving through groves of native trees. These were the last remaining belts of original forest on the island. Linking up with them was the new growth of younger trees and shrubs. In the thirty years that Pat had been working here, with herbivores like cows and sheep excluded from munching through saplings, and some human help, the forests have started to return, and with them the birds.

It was early afternoon when we dropped our supplies at Hakalau Field Station, a large, low structure that Pat had helped construct decades earlier. It sat like a sort of wooden space-station, on stilts over the uneven ground, with the central eating, researching, and lounging area connected by walkways to smaller bunk-bedded dorms. Attractive, stripey birds with black masks called nēnē waddled nearby. Found only in Hawai'i, they are the world's rarest goose and an emblem of hope. At one stage there were only thirty surviving in the wild, but thanks to heroic conservation work (including a captive breeding program by Sir Peter Scott, son of the famed Antarctic explorer Robert Falcon Scott), there were now over three thousand. We had been crammed into the car amidst cutlery, pillows, and other supplies for the building, and once we'd unloaded these, we pulled on boots and walked across a grassy clearing and down a path into the forest. Within a minute, we were absolutely soaked by an all-penetrating drizzle. I could just about make out distant little birds in flight, but my binoculars were so rain spattered I had to lick their lenses to try to get them to work. Pat pointed out to me the extremely rare avians hopping in distant branches as we walked. "That's an 'amakihi," he said. I nodded sagely, but it could have been a Mars bar for all I could tell.

We made our way downhill for an hour, soaked in the slanting rain and craning our necks at small fleeting shapes darting across the air between trees. Pat was identifying these birds not by sight, but by call. He had the air of a man fully at home, in his soggy element. He grinned and breathed in deeply. It was his favorite place on the island, he said, perhaps in the

Pat Hart hugs a tree.

world, despite his sadness at how likely it was these animals would go extinct, and with them, so many others. The forest was beautiful and the air so fresh that if you looked at a patch of bark, each appeared as a tiny forest of its own, with mosses and lichens so delicate and green they seemed to glow. Pat had been measuring many of the trees here for years, charting their growth, and he recorded the changing birdlife as they sang within the growing woods. He knew many of the trees individually, and had watched many of them grow for their entire lives.

At last, we caught a good look at a bird. It was an ʻiʻiwi, bright scarlet with a thick, curved beak a quarter of the length of its body. Its mate darted nearby. The ʻiʻiwi lifted its head and sang out. From somewhere else in the forest, another bird called. Pat looked even happier. "I don't know what that is," he said. But he pointed out that the box on the tree behind me probably did. It was green and plastic and the size of a child's lunchbox. In it was a microphone and a small computer. The forest, he said, was full of these—all listened over by electronic ears, recording day and night. As I had discovered, it is very hard to know what birds are around you, even if you are there to see them, but by listening with his machines, Pat could overcome this problem. He told me about new, cheap recorders called AudioMoths whose sensitivities are far beyond those of human ears, ranging from lower-than-human levels far up into the ultrasonic range. Powered by just three AA batteries, these recorders are resilient and tiny, with the device inside the boxes approximately the size of a two-centimeter-thick credit card. What was even more impressive was that they could be trained. Using computer algorithms, the audio sensors not only recorded the forest, but could be taught to identify different birdsongs and tell which birds were around. Other devices that Pat wanted to deploy could even be trained to identify the whine of the malarial mosquito. When these detected the sounds they'd been primed for, the AudioMoths would ping a message to the lab, hours away by car, instantly updating a map of what species were where. Conservationists could tell by text message sent from the sensor box when home ranges of rare birds were invaded by fatal mosquitoes, instead of by finding dead birds riddled with malaria. These were remote, permanent machine ears trained to listen for animals and to communicate and log what they found.

I took this in. When I left university, I went to work in a similar forest in Mauritius as a bird conservationist, tasked with trying to find endangered birds. I would choose an area of forest, wake up at 5:30 a.m., and head out in the dawn. There I sat silently for hours, hoping to hear the distinctive sounds made by the rare pink pigeons—sounds of them landing, bow-cooing to seduce one another, of their young begging for food. I listened, too, for the squawks of the Mauritius parakeet, the shrieking call of the Mauritius kestrel. All these birds were so rare that they'd had to be captive-bred back from near extinction—the pink pigeon had been down to just nine birds, the Mauritius kestrel to just one known breeding female. Every individual I could find counted in their species' survival. Sometimes I'd get lucky and actually see one, perhaps even identifying which individual it was by the colored rings we put on their legs. Often, I'd simply hear one, far out of sight. As these sounds were so easy to miss, I would regularly spend four hours hearing precisely nothing, and enter this in my log for that morning. Sometimes, for days, I would fail to hear a bird in the territory I was waiting in, despite knowing that they were probably there. In this hapless and ineffective manner, sitting under one tree a day, in one patch of forest, I attempted to chart the lives of whole populations of these critically endangered birds. Even though I'd been trained to pick out their calls, sometimes I was distracted, or ill. I couldn't be everywhere at once. But Pat's forest had its own ears.

We humans are natural and expert pattern recognizers: Our evolutionary success has come from finding patterns in the world and exploiting them. What berry to eat. What time of year it grows. What scary noise means it's time to find another cave to live in. What plants' leaves falling indicate it's time to skin some animals and prepare to dress more warmly. We interpret signs from the world around us, identifying and sharing trends, and when we get it right, we continue to survive. We carry these pattern recognition tools into our lives today: You recognize the discordant pattern of angry drunkenness in the man's voice on the night bus over the other passengers, and you get off a stop early. You recognize the blushing of that person you fancy when you make a joke, and you think, *Perhaps they like me, too.* We draw figurative patterns—stickmen—that our eyes recognize as people, and our brains concoct individuals that aren't there in the patterns of branches in a gloomy wood at night. Our noses detect scent patterns that say the toast is about to burn.

One way of looking at biologists is as pattern recognizers who focus on spotting repeated forms and behaviors in the living world. That was my job in Mauritius, to pick out the pattern of rare bird sounds amid a jungle of insects, wind noise, and my own breathing. Pat's boxes were better forest bird pattern recognizers than me. And, unlike me, they could be everywhere at once, forever. But how did it learn how to tell what sound was ʻiʻiwi and what was not? Pat said that he used something called machine learning, a field of computing where software is trained to find patterns in data. He said that it was early days, but what he could already do with machine learning was mind-boggling; he simply couldn't imagine what might soon be possible. This green lunchbox was fucking cool, I told him. I then realized that the green lunchbox had likely heard me, too, and somewhere recorded in this box's data was our conversation. Perhaps in the future some researcher would train the algorithm to find human speech in all the forest audio from the past few decades, and the machine-learning algorithm would highlight this audio sequence for them, and unpick my conversation with Pat—and maybe, years from now, I'd somehow find myself in trouble for swearing about the box.

At this point, Andre mentioned that every postdoc job going now was asking for biologists with machine-learning experience. I speculated to Andre that this must be the future, that machines would be deployed to do biology. He looked sad. "Will we still get to go out and do the science?" he asked plaintively. "I sure hope so. That's what we biologists want to do." We stood in the soft rain and watched the charming little ʻiʻiwi hop and chirp from one red bottlebrush flower to another. Andre had a point. A computer could observe this, but it could not (so far) appreciate it.

Pat then interrupted my thoughts to tell me about a student of his named Esther who had trained a computer to pick out ʻamakihi bird songs from the forest cacophony, a skill humans took many days to learn. And it saved time: Biologists could use the computer to home in on and confirm a bird was present in a patch of forest in one day, when normally it would take weeks. When there aren't many conservation days to go around, these machines could save species. *Save species*! With the information and codes for the algorithm available online, "everyone can use it," Esther told the *University of Hawai'i News*. The algorithm could be tweaked to detect *any* animal species.

Boom.

Unless you have craned your neck for day after day in a wet and slippery forest, unsure if what you heard was what you thought, knackered, worried that you'd gotten it wrong, that you'd wasted your time, gathered terrible data, your mind aware of the closing window for saving the species you are listening for from extinction, you might not realize what a big deal this is. For me, this was an extremely big deal. And not just for saving species, but also for understanding their communications. Trained to pick up different species, the machines, Pat explained, were highlighting subtle differences between the individual birds. Through his AudioMoths, Pat had discovered that the birds he was observing had different accents and dialects in different places, that individual birds had different voices, and they varied how and what they sang depending on their situation. Where previously a biologist could have said, at most, "This bird is here," the machines could help us see that each bird was different, that their calls changed over their lives and evolved according to the world around them. Because they were recorded, their songs could be compared and analyzed. Patterns in biological sound, shifting in place and time.

I asked if this made Pat think about what must have been lost, all the birdsong cultures that could have been here when the koa trees we'd seen were saplings, before the forests were cut down, when the birds flourished across the island at the time of Captain Cook's arrival. He looked down. "Yes," he said. "So much has been lost." We headed back to base leaving the little box on its tree, listening and waiting and sifting.

I can remember vividly the first time I swam in Hawai'i. I was on a sandy beach and walked out into the surf. As the next wave arrived and broke, I dove beneath it and swam out a little underwater. Immediately I heard the whales. I was so startled that I wondered if my mind was playing tricks

An ʻiʻiwi bird perched on a bottlebrush flower.

on me. The underwater soundscape I was used to was of breakers crashing and boats driving, my own breathing in a snorkel: loud, indistinct, messy sounds. This was something else. I could hear a number of whales, their songs overlapping one another. Some loud, some quiet. Grunts, squeaks, groans, bellows, and long, plaintive wails. When I pulled my head above the surface and into the air, they were gone, and I could hear once again the people on the beach, shrieking, playing, oblivious. It was like having been to a secret gig in the sea.

I'd been introduced to LOHE through a whale biologist named Marc Lammers. He used static listening devices like the AudioMoths called EARs (Ecological Acoustic Recorders) on the seafloor to listen to the whale songs and keep track of the humpbacks moving around the Hawaiian Islands. Until a few years ago, the main tool we had to count humpbacks was humans sitting on deck chairs on hills with binoculars and notepads. While this is undoubtedly still helpful and continues in practice, Marc's tech has been a game changer, allowing people to accurately count the whales as they arrived each year. He likened the change in his work from new technologies like these to going from "looking through a keyhole to looking through a porthole."

Many other residents of the islands had fallen in love with these mournful and mysterious sounds and, in 2003, some of a more technological bent had tried to find a way to constantly pick them up and relay them so anyone, anywhere around the world, could listen to them, live. They formed a nonprofit called the Jupiter Research Foundation (JRF) and called this the "whale-o-phone." But the coasts of Hawai'i are noisy underwater soundscapes, so to get away from the background of waves crashing and shrimp clicking their claws, they'd made some prototype stations off which to hang their hydrophones, far offshore. One thing led to another, and before they knew it, they'd designed a solar-powered, wave-propelled, whale-listening sea robot called a Wave Glider.

Bugging birds and bugging whales are similar jobs in some ways: In dense forests, as in the sea, vision is not of the utmost importance. Since both whales and birds communicate mainly with sounds, listening is often

Europa and Beth at JRF.

the only option. However, unlike forest birds on an island, whales migrate thousands of miles to places where recording devices cannot follow them. The Wave Gliders allowed biologists to go a step further and send robotic ears to voyage where humans could not easily go. I'd heard their development had recently stepped up a notch, with Wave Gliders enlisted to solve a pressing whale mystery: Marc's EARs, scattered off the shores of the main Hawaiian islands to count humpbacks, had registered a shocking silence. Most of the whales had disappeared. And so it was that I was invited to JRF to meet a robot called Europa.

Europa is what is known as an autonomous surface vehicle. She's made of two parts designed to harvest both information and power from the ocean. On the surface sits the "float" equipped with solar panels, a command-and-control unit, an audio payload, and transmitters, with a flag merrily announcing her presence to any ship that might run her over. Dangling twenty-six feet beneath the float is a submarine, which captures wave energy and houses a hydrophone recording continuous audio. With these research tools at her disposal, Europa bobs along on the surface of the sea, her propeller driving her at 1.5 knots (just under 2 mph). She can make way for twenty-four hours a day, for months at a time, recording away beneath the crashing of waves, wind, rain, and other surface noise. Europa can beam back her position, with photos from life onboard (seabirds perched atop her back, exhibitionist fishermen exposing themselves to the camera), as well as the songs of humpback whales recorded where no ships go for months at a time.

Beth Goodwin, who helped develop Wave Gliders like Europa, heads JRF's humpback whale work. An athletic woman in her early sixties with neck-length auburn hair, denim shorts, and a blue T-shirt with "whale detective" written on it, Beth has had a lifelong obsession with cetaceans. She told me her first word as a child had been "dolphin," her first job had been training dolphins at the theme park Six Flags over Texas, and her undergraduate thesis had been based on work in the dolphin aquarium at Steinhart.

A competitive swimmer, she'd bring her wetsuit to swim laps in the dolphin pool at night. She was impressed at how the animals learned from her, mimicking her tumble turns at the ends of the pool, copying her flips when she jumped into it. Years later when Beth returned to the aquarium the dolphins, upon seeing her, began doing tumble turns. The astonished curator told Beth they'd never done them before, but Beth knew they had, and that the animals had remembered her.

Trained as a marine biologist, she once ran a whale-watching company on Hawai'i, and now, as well as running the Hawai'i operations for Jupiter, she captains their research boat, the *May Maru*. Beth had invited me to visit her and watch the launch of a Wave Glider on a new expedition. But when I pulled up outside the organization's base in the grounds of a large house on the west coast, there were 60 mph winds blowing—a sign that an almighty storm was about to hit the Big Island. On the way over I'd seen a powerful gust tip the metal bucket of a truck off the back of its flatbed as it drove. People were being rescued from the sea, which was whipped into a white fury. I eyed the surrounding palm trees and their coconuts warily. Europa sat in front of a large garage, alongside the *May Maru*. Copper coated with solar panels on top and a sort of trolley underneath, holstered into which were the various cables and rudders which would hang beneath her at sea, she looked a little like a plastic, round-topped door laid flat. (One of these cables was later munched by a shark during near-shore testing.) On top was the shaft of a three-foot-long, rubber-topped antenna, fixed with cameras and her large red flag.

Looking at the Wave Glider, I felt strangely worried about her resilience in facing the tempestuous open ocean, but Beth assured me she had been built to stay at sea and weather storms that no manned ship could. She showed me maps of Europa's previous voyages: Trips east across half the Pacific to Baja California, 1,800 nautical miles away; there, she had recorded whales singing where no ships had ever documented them. Another voyage west to the Marshall Islands saw part of Europa malfunction. Beth had had to race out, hiring boats and scouring the high seas for her wayward mechanical offspring before finally safely bringing her robot home. As Beth described this search-and-rescue operation, it felt as though she was describing the rescue of a being, not a tool. It was clear that this machine meant a great deal to her.

Europa's upcoming mission was to the Northwest Hawaiian Islands, a distant chain of uninhabited isles and seamounts where people ventured so rarely that the native wildlife was very little understood. It was of vital importance: Over the past few years, 40 to 60 percent of Hawai'i's eight to twelve thousand humpback whales, which would normally visit on their migration, had totally disappeared, and the whales that were arriving were doing so later and later. If they were dead, this would be a devastating loss. Hawai'i is the breeding ground for a great proportion of all the northern hemisphere Pacific humpbacks; whales from subpopulations that feed from Russia in the west to Alaska and Canada in the east all swim to these benign waters every winter. Where had they gone? Europa's mission was to act as a mobile passive listening platform to see if she could track down the whales using their songs.

The day after I left, the storm passed and Europa departed immediately from the nearby harbor. Within a week, she had trundled her slow but unstoppable way to the Northwest Hawaiian Islands, beaming back fragments of humpback and minke whale songs—voices in the deep recorded by robots, sent from far away. The Wave Glider found so many whale songs that it supported the theory that the whales weren't dead, but had possibly shifted their migration patterns. One theory for this change was that they moved in response to a giant temperature spike in their Arctic feeding grounds caused by the heating of the seas. This area of extremely hot water contained far less oxygen and was called "the Blob." Within it, many of the plants and animals at the base of the food chain had died, and in turn caused the deaths of millions of seabirds and seals and marine animals. Fears that thousands of whales had died have now been assuaged by the return of the humpbacks in great numbers to the main islands since then. But the Blob did seem to disrupt their movements—and with more Blobs forecast as the climate crisis worsens, there are fears that the whales might not weather repeated disturbances.

I asked Beth how she went through her data. On a different voyage, this time from Hawai'i, east across half the Pacific all the way to Baja California, Europa had generated some five thousand hours of audio data, along with hundreds of images from the surface and underwater. To my astonishment Beth said she and her human team went through it themselves, visually and aurally, scanning it three times over. It took four of them six weeks, eight hours a day, to manually pinpoint five thousand calls of humpbacks and other cetaceans from other noises. I asked if the work made her go mad. "We did. You start hearing things and seeing things." Beth now hopes to use machine-learning algorithms to automatically detect whales in the audio.

On the drive back from Beth's, my mind fizzed with all that I'd seen. Microphones fixed in the deep and whole forests rigged up, listening. Robots sailing solo voyages across entire oceans, powered by waves and sun, able to avoid ships, weathering storms, sampling the water, constantly beaming their findings home. I couldn't think of anything that fit Brautigan's poetic dream better—the endangered birds and whales and their listening machines of loving grace.

I made my way back to the hotel, and that night the storm calmed. The next day, my wife, Annie, and I watched the sunset. Tourists lined up for selfies, but on the ocean, unnoticed by all of them, it seemed, we saw two whales spout. They were moving north, about half a mile apart, and headed past the harbor at Kawaihae, beneath the Hawaiian war temple of Puʻukoholā Heiau ("temple on the hill of the whale") where, two hundred years before, King Kamehameha had dedicated the bodies of his slaughtered rivals to his gods before heading on to take control of the rest of the islands. I thought how strange it was to be able to see something as intimate as an animal breathe from so far away.

These powerful new tools were all invented in the last decade. They trailed humans playing catch-up with their vast recordings. Beth had described how overwhelmed she was trying to listen to all the recordings from each machine voyage. There was too much data. Across the world's seas and forests, more of these machines are being launched and placed, with ever more ways of listening. This snowballing of information for humans to sort through had seemed unsustainable to me—and yet Pat's point about our ability to train machines to comb through this data raised an intriguing question that I was desperate to answer. It was one thing to build a machine to record and identify the call of a whale. It was quite another to train one to try to find any meaning in those calls. Could this, too, be done?

Could these pattern-finding machines be what we need to unlock the mysteries of animal communication?

9

Animalgorithms

Machines take me by surprise with great frequency.
—*Alan Turing*

From the Big Bang until 1877, if something made a noise you either heard it or you didn't. But Thomas Edison figured out that you could engrave sound—itself made up of vibrations in the air—onto a series of pits on a sheet of tinfoil. You could give the sound a lasting imprint. Then you could run a stylus along these pits to bring the vibrations to life again. At first, humans simply recorded other humans, but some soon turned to recording nature. On May 18, 1929, Arthur Allen, an ornithologist at Cornell University, set off for Renwick Park on the edge of Cayuga Lake in Ithaca, New York. He took with him technicians from the Fox-Case Movietone Corporation. They set up remote microphones by a branch where Allen knew a song sparrow would come to perch and sing, and they waited. It came, it sang, and they made a unique recording. It was one of the first nonhuman voices ever recorded. A few years later, Allen led an expedition to Louisiana to search for ivory-billed woodpeckers and managed to record their calls. Later the woodpeckers disappeared, thought extinct, but their voices remain.

At first, after sound recording devices were invented, you couldn't really compare the recordings much, apart from listening to them.

J. J. Kuhn, a local guide, and Peter Paul Kellogg, Cornell University, making a sound recording of ivory-billed woodpeckers in the Singer Tract, Louisiana, 1935.

But in the 1950s, people devised ways to make pictures of the vibrations. These were called sound spectrograms. Like a musical score, a spectrogram represents time running from left to right, and the frequency (or pitch) vertically; the color or brightness of the lines is used to show the signal strength. This is sound made visible. Frozen in time, you could now not only listen to your recording as many times as you wanted, but you could also *look* at how two or more sounds varied, and measure this. Humans aren't very good at listening to two things at once, but our eyes are really good at hunting differences, comparing and measuring. By converting sounds into pictures, the science of finding patterns in them was suddenly a lot easier. Here is what the spectrogram of a pod of killer whales calling at the same time looks like. It is pretty nuts-looking. All these lines are different whistles and buzzes made by the whales talking to each other.

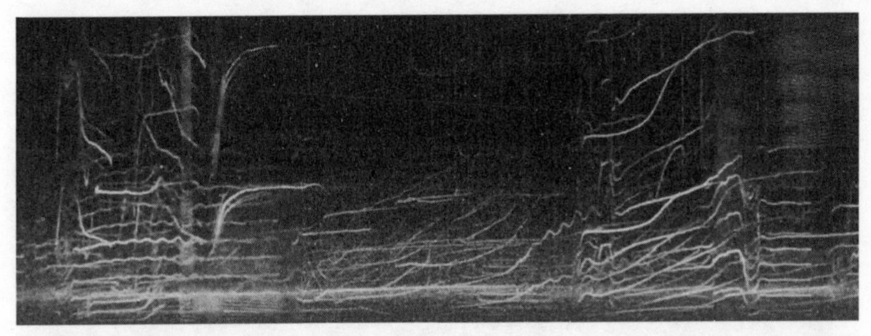

Recorded by biologist Jörg Rychen, this was the winner of the IBAC "Craziest Spectrogram" award, 2019.

As recording devices became portable, naturalists could bring back sounds from across the globe: the calls of gibbons, birds of paradise, cicadas, and whales. We could now archive, analyze, and compare living sounds. We could amplify them and play them back to animals to see what they did in response. We could use synthesizers to generate new sounds and make microphones capable of recording sounds we cannot hear—the low infrasound noises of elephants, the high-pitched chirps of bats. We developed hydrophones to work underwater, where the changed behavior of sound confounds our hearing. These inventions ushered in a fresh field of science, bioacoustics, helping us to study the sounds of life.

It has become possible to send sound recording devices to places we can't go, from the Bermudan seas, where Roger Payne's singing humpbacks were first perceived, to the pools of Diana's dolphinariums, to dangling beneath Beth's autonomous sea vehicles, and clamped to trees in Pat's forest reserve. For those who dreamed of understanding animal communications, recording their voices was a monumental step forward. But, as I was learning, it was also a Pandora's box. For what do you do next with all these recordings? Pat had mentioned that you could train computers to sift through them, giving him vital conservation information on which birds were singing, when and where. But I'd also heard of researchers using AI to find other patterns within acoustic data—to begin to decode not just who was speaking, but what they might be saying.

* * *

An early machine-aided listening device.

The International Bioacoustics Society (IBAC) was founded in Denmark in 1969 to gather all sorts of people together from archivists to animal behavior scientists in "informal settings" to share their discoveries and ideas. In August 2019, I was delighted to learn that their annual conference was taking place nearby, at the University of Sussex, and I signed up to join in the fun. I drove through the university campus, looking for the set of rooms where the conference was to be held. It was late summer, and the place was mainly deserted. Seagulls squawked and flocks of rooks swooped through clusters of glass and brick buildings, concrete walkways, and lawns. I checked in to the conference and was given a bag, a mug, and a schedule. I learned that the "informal settings" would include five different themed pub crawls, a visit to a stately home and deer park, an animal sound recordings audiovisual DJ concert, a gala dinner, and a demonstration of Morris dancing. There were awards for the posters, the presentations, and the "coolest animal sound imitation" (these people were *really* good at making animal noises). There was wine and beer and cheese and coffee in rotation. But most of all, day after day, for six days, in twenty-minute slots from 9 a.m. to 6 p.m., the people who record and analyze the sounds animals make stood at the front of an amphitheater, played those animal sounds, and talked about what they might mean.

Over the next few days, I was startled to discover that in labs and farms and dusty plains and tropical swamps across the world, people were using sound recording and manipulation devices to play dogs the cries of distressed babies and humans the cries of distressed dogs; to record rats given MDMA (the psychoactive chemical also known as ecstasy) squeaking with "happiness"; to record the grunts of piglets waiting to be reunited with their friends; and to play an elephant's trumpeting through huge subwoofers to other elephants in order to make them think another big elephant was there. Tiny jumping spiders danced and made noises by vibrating their bodies in dazzling audiovisual mating displays. In Sweden, a group of scientists had placed microphones on cats and were following them around with cameras on their own heads, filming cats wanting food, cats trying to get through doors, cats upset at being picked up, cats happily being stroked, to see how their different meows and purrs and wails might signal their feelings and intentions.

I learned about crying and voices, and how the unique shape of the larynx gives every single mammal a unique and identifying voiceprint, like a fingerprint. This acoustic fingerprint means you don't need to say your name for others to know it's you when you speak. You develop your voiceprint as soon as you use your voice—and two days after birth, many mothers, be they fur seals or humans, can recognize their offspring by their voice alone. Computers had now been trained to recognize them, too. As wonderful as the IBAC was, however, I must admit that there were some strange moments. Sitting in a lavatory stall at the break, with the cubicles alongside me full of some of the most attentive, analytic listeners on the planet, I was paralyzed. What would they deduce from listening to my lavatory melody alone? I'd never felt so acoustically exposed.

The first thing that struck me at the conference was the vastness and complexity of the world of animal sounds, and how our assumptions have been wrong about so much of it. Everywhere we chose to put our mechanical ears, new animal communication behaviors were discovered. In the lakes and rivers of France, a woman discovered 271 species that made sounds underwater, where no human had ever heard them before. She talked of the birth of a new field, "ecoacoustics," where you listen to the *biophony*—not a single animal noise but the interacting and overlapping sounds of an *entire living ecosystem*, from macaws to frogs to beetles. There were levels of complexity in animal sounds that surprised those who found them. Budgies, for example, used "vowels" and "consonants." Cats had large "vocabularies" of different calls. The cries of pigs could show us how they felt, and machines can be used to automatically track their happiness by listening to their voices.

Old assumptions were being overturned, and a striking example of this was birdsong. Many birds are fantastic singers. I'd thought, as perhaps you do, that male birds were the main singers, as had Darwin and most naturalists. It turned out that if you listened to lots of different birds, this was wrong. At the conference, a team led by Katharina Riebel from Leiden University shared the extraordinary results that in their analysis of all songbirds, they found that 71 percent of species have female song, and females sang in every group of birds. Katharina was left "speechless" by the discovery. The only conclusion that made sense was that female birds now, and their ancestors before them, were singers. So why did we assume that singing was a male thing?

It turns out that in the temperate northern hemisphere, where Darwin and many early ornithologists were from, the males sing while the females are generally quiet and drab. This was assumed to be the case everywhere in the world for all birds, and singing was deemed a largely male pursuit. As Western ornithologists spread across the world, they took with them their prejudice, and when female birds were sometimes recorded singing in the tropics, they were considered strange outliers. Now that (mainly female) scientists have finally started to investigate female birdsong properly, they have discovered that female birds do in fact sing in the northern hemisphere, too—just more quietly and less often. It was easy to miss, especially if you weren't listening for them. As Dr. Evangeline Rose, a biologist from the University of Maryland who studies female song, told *Psychology Today*, "There's been almost a century and a half of research on male song, while studies on female song only started in earnest in the 1980s." She thinks in missing female song, where singing has more diverse functions, we have overlooked a more complex story than in the song of males.

I was astonished by this. If something as basic in animal sound was so misunderstood in species that we have scrutinized closely, what more was there to discover? What could this mean for humpback whale song? The main assumption for its function was borrowed from what we assumed about birdsong—namely, that it was a way for male humpbacks to show off and get laid. But could this be just part of a more complex picture? It's hard to tell which whale is singing, let alone determine its sex (you try to take a look at their genital slits, in case you were wondering). It's assumed that when you hear a humpback song, a male is making it, but what if that isn't true everywhere, or all the time? What might their songs mean then?

I thought about how the loudest whale sounds tended to be made by the males but the most cooperative, long-lasting sociable groups of cetaceans tended to be female. Were we missing the more interesting conversations by starting with the shouting? The fungal biologist Merlin Sheldrake talks about how helpful queer theory, which explores nonbinary ways of dealing with identity, can be to biologists: "If you don't presume to know what this organism is before you start investigating it—if the very nature of its being is a question—then you get to some interesting places."

The second thing I noticed at the conference was what I had wondered about in Hawai'i—that the ability to record unlimited amounts of sound, paired with the limited life span of humans, was proving troublesome. During the conference, I'd watched scientist after scientist play their recorded sounds, and then show their spectrograms and the statistical analyses they'd peformed. To conduct these analyses, they had had to go through their sounds using clunky computer programs and label where they started and ended. Sound by sound, they'd classify each as one type or another and save them, then run processes to clean them, then arrange them into databases and label and organize these. The programs they used were unintuitive and ugly. The work was drudgery.

Big data is the dream of many computer scientists—the more data there are, the more patterns you can locate in them and the more powerfully you can train algorithms to find, classify, replicate, and otherwise mine that data. But the big animal data had become too big for all the poor biologists; many seemed almost trapped by the process of segmenting (marking where a sound began or ended), labeling, organizing, cleaning, representing, and analyzing their recordings. I wondered if they ever got much time to do what Roger had done and just lie down and close their eyes and listen.

But thankfully, computers could help sort out their prolific digital offspring. One young man, a flame-haired, goateed, unflappable Kiwi named Wesley Webb, was so fed up with the tiresome business of dealing with his thousand-plus recordings of New Zealand bellbirds that he'd teamed up with a data scientist named Yukio Fukuzawa to make a program to do it for him. "Koe" could sort all your sounds, in bulk, automatically by their acoustic characteristics, and arrange them in a giant visual cloud of individual sounds. You could ear-test any of the sounds, select entire clumps and label them, ask the cloud to rearrange them and color-code whole groups of them, then sort and re-sort them. It could analyze them for you, too. Normally you must do this work sound file by sound file, but this was an intuitive and web-based program, meaning lots of untrained people could work on the same database across the world simultaneously. It was free for anyone to use, he told the conference room, before continuing to present his PhD research on New Zealand bellbird song cultures and explaining how Koe had massively sped up his classifying and measuring of the 21,500 song units in the recordings he'd made, saving him months of time. At the lunchtime demonstration of Koe, the hall was packed with scientists asking if it would work for bats, frogs, or dogs (yes, yes, and yes). Wes won the prize for the best presentation.

I got the impression that a major bottleneck was being removed, untold human hours were being saved, but also something more was going on. If you have lots of recordings, and many of them have been labeled and organized, then not only are you learning about whatever animal you are recording, but the machine you are training to sort and process can learn, too. This was why I had come to the conference. Artificial intelligence had already played a part in my journey, and I suspected it was going to be transformative in what came next.

Allow me to return to the origin of this story. As that humpback whale whose life had so dramatically intersected mine propelled itself from the sea and erupted into the sunlight as countless millions of its forefathers had done before, it was immortalized as none of them ever had been. The arc of its flight was recorded by a man called Larry Plants on his cell phone, and also by a woman filming from the shore, and a ship's captain snapping photographs. All amateur whale photographers. They then uploaded their footage to the internet, where I could find it. Their GPS position (and therefore the whale's) was automatically logged, the time of the breach automatically stamped onto the videos and photographs. As it crashed back down, it left behind an indelible digital footprint. Deep below, on the seafloor, an underwater microphone set up a couple of weeks before recorded the sound of the crash it made as it landed. From above, satellites took myriad photographs, charted the weather, the surface temperature, and other readings.

That day, like every day in Monterey Bay, thousands of photographs had been taken on the water: holiday snaps from whale watchers and the crews of the vessels. Normally these would all live in the private collections of the photographers and never be seen again. But in a stroke of luck, just a fortnight before that violent morning, a whale researcher named Ted Cheeseman had set up a website called Happywhale. Ted, a lean outdoorsman with close-cropped dark hair and a boisterous puppy, realized that whale watchers were a massive, free, and global whale-surveillance network. He gave them a platform on which they could upload their pictures—and specifically their pictures of the whales' tails. Because tails are vital if you want to know about your whale.

OK, so they're not technically "tails." The long muscular rear trunk that propels a humpback whale is actually called its peduncle. As wide as an ancient oak where it leaves the whale's pelvis, it tapers to a neat joint with a huge double-sided paddle, which is called a fluke. Each whale's fluke is different. In color, the splotchy patterns of light and dark pigment vary between the different tribes of whales, from Antarctic to Arctic, Tasman to Newfoundland. Across this flesh canvas is engraved its life story, as the scars on a chef's hands attest to paring knives and oven doors. Killer whales attempt to drown baby humpbacks by pulling them down by their flukes, so many humpback whales carry chomp marks from killers' teeth, the scars stretching as the whales grow to adult size. Barnacles leave constellations of ringed scars, cookiecutter sharks chew out divots of flesh, boat propellers slice sickle tracks, and snagged fishing lines carve telltale cheese-wire slices. A tail is both a fingerprint and a flag. As a humpback dives it will often raise its fluke out of the water, drawing gasps of human admiration and the clatter of digital camera shutters.

For decades scientists have identified whales by these tail flukes; human researchers spent tens of thousands of hours staring at piles of photographs at the end of a season's whale expeditions, matching stacks of similar flukes to deduce who they belonged to. From this they could map where the whales photographed had traveled, who they were with, what they were doing, who they had given birth to, and how old they might be. It was grueling work, requiring intense attention to detail, and generated frequent mistakes.

Ted had been given more than 150,000 humpback fluke photographs and now has over half a million. By combining all these citizens' photographs with the existing fluke libraries, he turbocharged the scientists' databases, like Interpol pooling fingerprints from crime scenes all around the world.

At Happywhale, Ted upgraded his system from biological fluke matchers—people—to computers. He and his team took 28,000 labeled photos and 5,000 photos of unidentified whales and, with a $25,000 reward from Google, he challenged anyone to write a computer program that could match the unidentified whales. A staggering 2,100 different teams entered. One of the winners of the competition was Jinmo Park, a computer scientist from Korea. This man, who had never seen a whale before, trained a computer vision AI tool (called a Densely Connected Neural Network) using the photos Ted provided. Neural networks are information-processing software tools modeled on the networks of neurons in the human brain. They are very good at finding patterns in data, and are the backbone of the branch of AI called machine learning.

The deep-learning algorithm that Jinmo Park wrote processed the 5,000 photos of unknown whales and correctly identified 90 percent of them. Ted and the programmer behind Happywhale, Ken Southerland, took the algorithm Park had programmed and began to feed it images of other whales that neither he nor any of the other expert human whale identifiers could recognize. They were tricky photos to ID, like whales with all black or all white tails, or whales in fuzzy photographs. Ted told me that he didn't think this was a big deal, but it turned into one of the most significant moments of his life. The computer started matching whales that no human had ever matched. He didn't believe it. So he checked himself, by eye, looking at photographs he wouldn't have matched until his attention was drawn to their similarities. The algorithm was right.

Every week, Ted added thousands of images to the Happywhale memory as data from across the world poured into his "fully automated, high accuracy, photo-identification matching system." With superhuman focus and access to terabytes of humpback data, the algorithms compared and learned and found new patterns—crucially, patterns people had missed. Newly digitized archives from decades past linked old black-and-white shots of calves to today's living middle-aged whales, filling in their backstories. Affinities also became apparent. The algorithms found that some whales seemed to be consistently spotted with other individual whales, turning up again and again together from one sea to the next, year after year. They were whale friends, traveling thousands of miles to feed and sing with each other. Families were mapped and journeys traced; the algorithms linked previously unrelated sightings across entire oceans, connecting whales in Japan to sightings in Russia, Hawai'i to Alaska, Antarctica to Australia.

For those who uploaded their snaps and could now identify the animals they'd seen, these creatures became less anonymous beasts of the sea and more individuals with inclinations, histories, and friendships, with algorithms now helping to connect their life stories. The more they knew about them, the more whale watchers felt bonded to these animals, and they anxiously waited for the return of their favorites from their breeding grounds. I met a man whose wife had died; the whale watchers had named one of the whales after her, and now he went out many times a week, hundreds of days a year, to see if she would return. He tracked her on Happywhale. He told me how one day she had returned safely from the breeding waters with a new calf. She had breached and he had seen her eye, he said, with tears in his.

And so, three years after our near death, I had asked if Ted could use the footage and photos recorded by the whale watchers active in Monterey that day to identify the whale that leapt on us. And he had. Or rather, he and his algorithm had.

This is the whale.

And this is its tail (from a photo taken minutes before the breach, with our kayak tour coming into view!).

Its number was CRC-12564. Ted looked up its record and linked it to sightings elsewhere. I learned it was born seven years before it breached on us, in the waters of Central America, and who its mother was. Photos in Ted's database offered glimpses of it feeding, socializing, breaching in Californian and Mexican seas. Detailed photos of its body showed the kind of scars that reveal it had been caught in and then escaped from fishing nets, and other kinds of scars hinted that it might be male. The whale had returned to Monterey each summer after it leapt onto me, but it hadn't been seen for a year. I signed up to "follow" the whale (which Ted had named Prime Suspect), and a few months later I received an automated email, letting me know it had been spotted (by a human photographer, plus a machine pattern recognizer) safely again. The more I learned about this whale, the more it wasn't just "a whale"; it was an individual. I felt connected to it. I cared about it. I wanted it to be OK.

I was gobsmacked. A whale lands on you and disappears. End of story. But thanks to lots of people who liked looking at whales and their intelligent machines, it was not the end at all. Machine learning and other branches of AI influence our daily lives in myriad ways. They have helped this book come into existence, with an algorithm transcribing the hundreds of hours of interviews I conducted for it. Other algorithms have checked my spelling and finished my sentences for me as I typed them. Google's effective prediction of my email responses has made me realize how predictable a lot of my writing is (sorry, reader), and by extension, perhaps, most human language is. It has saved me huge amounts of time, and I have ended up spending this saved time procrastinating by looking at my phone, at news apps and shopping sites and social media, all of which have been beautifully designed and pumped full of AI whose purpose is to suck up my time and money and data.

AIs are used to look at MRIs and find tumors; they're used by engineers to look at national grids and direct power across the country; they're used to test the limits of human ingenuity by beating human players at chess, Go, and video games, to examine videos of animals I've taken in bad conditions and enhance them to look better. They peruse our digital presence and bank statements to decide our credit ratings, and scan documents in Chinese and English to translate between these languages. These AIs, like all the ones we have made so far, are "narrow," meaning they have only one or a few specific tasks. Of course, they don't have a concept of *what* they are doing. They don't know breast cancer is bad, that winning a chess game is a triumph, that an image is beautiful, that stopping blackouts means homes have light, that buying this house means I can grow vegetables in the garden, that the end of this sentence is important to me. But they can already do all these things, faster and often better than we can.

In biology, AI has been used to discover that male mice sing different songs as they court and have specific calls for playing and anticipating tasty food, and when they are upset. Another team used trained computer vision to analyze the fleeting expressions on the faces of mice and linked them to how they were feeling, leading them to discover that mice have at least "six basic emotions."

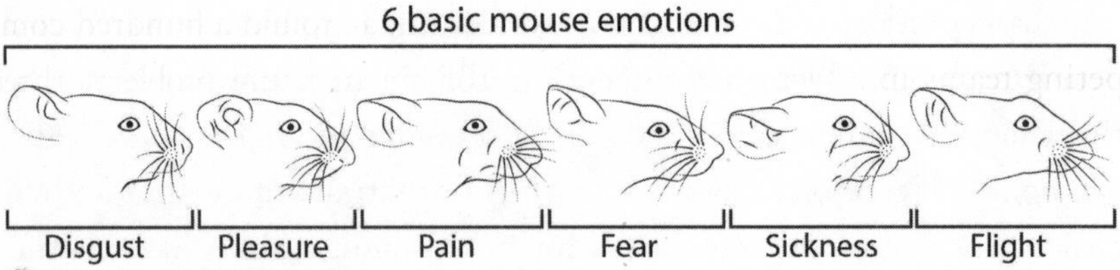

The six "basic mouse emotions" discerned by AI.

Planes flying across the Arctic have AIs scanning their camera systems to spot sleeping polar bears beneath the snow. Some scientists using AI have found "arguments" between Egyptian fruit bats were different when about food or resting spots; AIs have scoured satellite photos and discovered hundreds of millions of trees in the Sahara Desert where there were previously thought to be none, and predicted volcanic eruptions days before we could. Since I met Ted, Happywhale has moved on from identifying humpbacks exclusively and can now ID individuals from more than twenty species of whale, and from pictures of other parts of their bodies, not just their flukes.

Machine vision is now being used by biologists across the globe. For example, the AI nonprofit WILDME has developed open-source platforms for fifty-three other species and counting: manta rays, giant sea bass, skunks, and sea dragons. And the data continues to pour in—at the Monterey Bay Aquarium Research Institute a deep-sea database called

FathomNet is being made publicly available with twenty-six thousand hours of deep-sea video, a million images, and 6.5 million human annotations. Machines can be trained to find patterns in other scientists' work as well. In 2021, AI was used to carry out a meta-analysis of one hundred thousand climate change studies, carrying out important drudgery outside the scope of human capability.

In November 2020, the biochemistry world was rocked by something called AlphaFold. This was the project name for deep learning software developed by DeepMind, the Google/Alphabet–owned AI company whose stated mission is "solving intelligence, and then using that to solve everything else." AlphaFold made what the journal *Nature* described as a "gargantuan leap" in solving a long-standing problem in biochemistry—figuring out how a protein would fold. AlphaFold crushed around a hundred competing teams in a biennial competition, solving structure problems three times more accurately than the winner in 2014, and far faster. The program is so good that one researcher, Mohammed AlQuriashi from Columbia University, predicted many chemists would simply leave the field of protein structure prediction as "the core problem has arguably been solved." Solving this problem is fundamental to disentangling how our cells work, and has implications that will affect our lives—in the creation of medicines, the understanding of aging, in bioengineering. "It's a game changer," said Dr. Andrei Lupas, the director of the Department of Protein Evolution at the Max Planck Institute of Developmental Biology. And because machine learning is so general purpose, many tools developed in one field can often be easily adapted to transform another.

If you're wondering *how* they do this, think of AIs like human children: They are hungry for information. When you teach a little child how to speak, you do not sit them down with a book of syntax and grammar. You speak to them, a lot. The toddler mimics you, emulating the data you have given. They speak back to you, but if they say something wrong or inappropriate, you do not generally tell them the principles of speech, you often just tell them the correct sentence for that situation, and you wait to see if they can correctly replicate what you have said in the right context.

This is called reinforcement. The child's brain does the rest: It remembers the situation and next time tries again, perhaps with a new variable, until the sentence output is correct. This is, of course, an oversimplification, and there are many kinds of different AI techniques used in the examples above. But whichever type is used, and however it is trained, a computer brain focused on one task can do this again and again, day and night, a lot faster than a human brain, forever. As with human brains, if you ask that toddler how it knows to use the correct word, they may find it hard to explain. Similarly, the exact workings of the AI can be inscrutable, but somehow it *has* learned, and if you train it well and feed it lots of data, it can get it right. And once it is good at getting it right, you can set it to work on huge data sets, beyond the scope of any one person to process. In the words of my friend and AI expert Ian Hogarth, this technology is a "force multiplier."

So, what could machine learning and other forms of AI discover in the utterances of cetaceans?

Project Tidal's AI fish behavior recognition system (used to identify disease and fish feeding patterns).

On the final day of the conference, the entire morning was devoted to the discussion of whales and dolphins. Cetaceans, as I'd learned, seem to use something like names; at the group level, sperm whale and killer whale analysis indicates they also might have sounds for their social tribes. Scientists studying signature whistles have to sift through recordings of dolphins communicating to find them, looking through the spectrograms for telltale shapes. And finding these signals is really hard: Dolphins are chatterboxes, and they can, en masse, produce vast quantities of whistles and sounds at the same time. One scientist, Jack Fearey, played a recording of wild, common dolphins off South Africa, vocalizing in a group of a thousand. When these huge pods move at speed, it's called a stampede. The sound underwater was an astonishing wall of interwoven whizzes and whistles and buzzes, which he described as "a dolphin cocktail party." It is very hard and labor-intensive for humans to pick out dolphins calling their own and each other's "names" in the recordings of such an enormous gathering. But he gave it a go, searching through all the spectrograms by eye to find ones that looked like they had the features of signature whistles. He found 497 whistles in the recordings and, out of these, twenty-nine looked like signature whistles. When he used a computer analysis, it encouragingly came to the same conclusion. Now that he knew the computer analysis worked, he could be more ambitious and scale up to recording massive data sets that no human could comb through by eye to look for signature whistles. Jack announced his next plan was to rig the Namibian seafloor to record years of continuous audio and to try to find all the names of all the dolphins at the party. But both the human and computer analyses were held back by the same problem: In a "cluttered acoustic environment" such as a dolphin stampede (or a human cocktail party), huge numbers of the signature whistles and other sounds were masked by the animals talking on top of each other.

The person whose talk I was most eager to hear was Julie Oswald, a Canadian in her forties with short brown hair who works at the University of St. Andrews. She'd grown up in Kitchener, near Toronto, "nowhere near a dolphin," and started out as a nurse—but was drawn to dolphins. She'd moved into bioacoustics, frustrated with how hard it was to measure anything else in behavioral science. At last, here was something that seemed quantitative; you could put it on a graph and compare it! Signature whistles were a logical first discovery, a "word" that individual dolphins make that always sounds the same. (I wondered, if we could figure out the dolphin whistle for "human" and combined it with pointing, could we then have the first, tiny, provably meaningful interspecies conversation? "Me, human; you, dolphin"?)

Julie's was the final presentation of the conference, and it was worth the wait. She explained that dolphins make lots of noises besides their echolocation and the easily recognizable signature whistles. They also produce other kinds of whistles and "burst pulses," which are a series of rapid clicks. Many of these are inaudible to humans, meaning that some of these dolphin communications have been detectable only recently. We do not know the full variety of sounds a dolphin makes, or how much those sounds change through a dolphin's life, let alone how much the vocalizations differ between individuals of one dolphin species or between species. So Julie has started to explore this new sonic world in order to identify patterns. To do this, she recorded a group of thirteen captive dolphins in an oceanarium in Spain, twenty-four hours a day for two months. She then took the 1,500-odd hours of data and organized them, first running it through a program that could extract the whistles from the other sounds, then another program to clean it up—a dynamic time-warping tool to make the whistles the same duration for easier comparison. Finally, she ran it all through an "unsupervised neural network" to find out how many whistles were in the recordings. This is a kind of machine-learning tool that, like many others, is based on artificial neural networks. Ted used neural networks in Happywhale, for instance, to match humpback whale tails. But Julie's neural networks were of a kind where the computer doesn't get any help from a human to label and score the data it is given (hence, "unsupervised"). This was very different from how animal acoustic analysis had been done until recently. Gone were the days of recording audio of some dolphins, printing out spectrograms of the sounds they made, and visually looking through these, manually highlighting the bits that you thought looked different.

Julie's AI extracted 2,662 individual whistles of 342 consistent types. That was a big range of different kinds of sounds. She wanted to know how many more kinds of signal she'd have found if she'd recorded longer. If you listened to some people talk and wanted to count the numbers of words they were using, the words you'd hear at the beginning would be completely new. If you plotted the total number of new words on a graph against a period of time, it would spike with your most commonly used words: your name, conjunctions such as "and" and "the," and common words such as "please" and "thanks." Then it would start slowly leveling off with words you used more rarely, like "tree" or "breakfast," before near flattening, with occasional new words that you might not use often at all, like "funeral" or "bikini." Even at the end of Julie's 2-month study listening to dolphins, new special whistles were being discovered at around the rate of 1 a day. She extrapolated that the dolphins had a repertoire of around 565 whistle types.

I couldn't believe what I was hearing. Dolphins could have repertoires of more than five hundred whistles! Julie found similarly impressive results with recordings of wild dolphins. For an acoustic signal, like a spoken word in humans, to have meaning, it has to remain stable. We wouldn't be able to understand one another if we kept changing the words we used. So next, Julie planned to check how the different whistles were used over time, and if the dolphin whistle repertoire was stable, like human words. This is not to say that dolphins have words. What these acoustic units she discovered mean, if anything, is completely unknown. But if you recorded humans communicating and broke up the acoustic units into types, you'd have a graph very similar to Julie's. Alone, Julie could have done none of this. Computers had sensed, recorded, represented, organized, cleaned, compiled, and analyzed dolphin whistles that no human could hear, and within them discovered patterns no human could perceive. Later, I asked Julie if you could call her findings a dolphin vocabulary. She said no, that if "vocabulary" was used instead of "repertoire," it might make you think that the whistles had meaning and syntax. But if the first whistle to be decoded did have meaning, she said, then perhaps one day that's what we'll call it.

I wanted to fast-forward to that day.

One of the excursions at IBAC was to a nearby stately home called Petworth House. Along with a hundred bioacoustics researchers, I wandered the historic pile and grounds. As they walked through a wood-paneled antechamber, two of the scientists started calling out different notes,

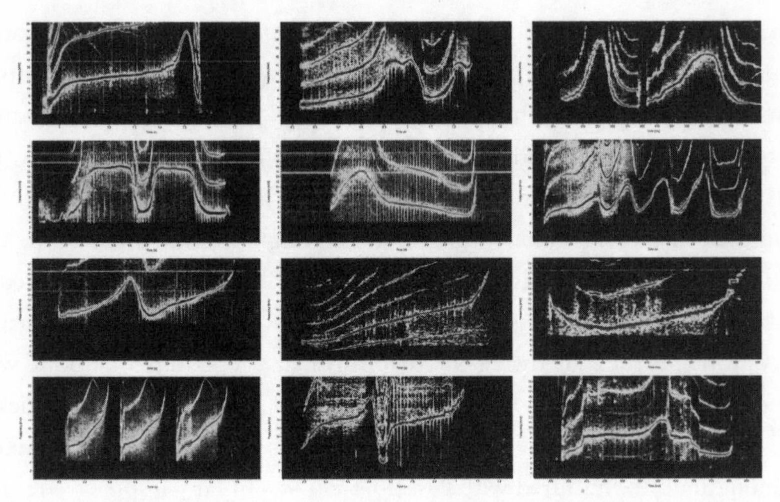

A small sample of the hundreds of different whistle types that dolphins produce. These are from bottlenose dolphin whistles (courtesy of Vincent Janik, University of St Andrews).

singing to discover the dominant frequency of the hall. Henry VIII and other Tudor royalty gazed down from the walls at humans who'd come here from lands that the Tudors had never heard of. In a far room, I found an ancient globe, one of the earliest terrestrial globes ever created. It was crafted by Emery Molyneux in 1592, and centuries of probing fingers had worn England almost entirely off its surface. Careful mathematical filigrees crossed known continents, and those newly discovered, tracing the journeys of Francis Drake and the outline of a place called California. At the time of the map's drawing, Europeans had not come across the continent of Australasia, though they had started to estimate the surface area of the Earth and had an inkling something was out there. Molyneux had drawn a fearsome whale-monster to fill the gap in the map.

Drake and his adventurous compatriots had no idea that in this void was a continent, and on it lived humans with diverse and dizzying cultures who had been there far longer than men had lived in England. I thought of how incredibly young our piecemeal mapping of the sounds one species of dolphin makes is, of the hundreds of whistles that Julie's algorithm had newly found, of the communication terra nova awaiting discovery.

Some of the cetacean map has now been filled in. I, like most other humans, no longer think that whales are just big, stupid fish, suitable only for industrial butchering. We have learned that cetaceans are mammals like us. They live long lives in complex societies where they cooperate using sound communications. They have clans and cultures delineated by the way they speak. I had come to comprehend their virtuoso ability to create, shape, transmit, and listen to sound. I'd seen their brains, whose combined features may hint at "higher" capacities like our own, and learned how experiments in captivity have already confirmed some of these, with small cetaceans like dolphins surpassing some of the cognitive capacities of our near relatives on the land—the great apes. They did things like we do, copying movements and mimicking sounds, following gazes, playing, recognizing themselves in mirrors. They bonded with their friends by touching their pec fins together, as we do by holding hands. They sang, they learned, they varied. They did things we consider altruistic, such as helping others in trouble, as well as things we consider evil, such as rape and the killing of infants. They were interested in new things, and in us. They were complex beasts. How much we now knew, considering that we used to assume them to be unthinking and incommunicative! How much more could we learn? Limited by our own senses, bodies, and brains, we are now augmented by machines that can voyage, listen, and begin to decode their lives for us.

But I'd learned something else—that although so many, like me, had become tantalized by the prospects of what we could discover in the communications of other species, it was not a priority. Scientists told me that their institutions didn't have the firepower, the willpower, or the money to go for pure research into mysterious questions. Grants had to be given for obvious conservation goals, or fisheries management, or finding ways for the Navy not to kill too many whales. Ted Cheeseman of Happywhale described biologists as always "a day late and a dollar short." Many didn't want to try to suggest research in this area in case they were made to look foolish, risking their careers and other funding. Others just didn't believe there was anything to find.

But with time running out for much of life on Earth, I worried that our recordings of whales could become all that remained of these unique animals' cultures, a pile of digital ghosts. Studying cetacean communications was extremely complex and ill funded, and made huge demands of researchers' lives. I developed both a deep respect for the difficult work behind what we'd learned, while also wondering if we ever stood a chance of learning to speak whale.

How, then, could we move further, faster? It seemed it would take something much bigger. A change of gear.

10

Machines of Loving Grace

To move forward is to concoct new patterns of thought.
—*Edward O. Wilson*

When van Leeuwenhoek looked into pond water, he discovered a microcosmos of "animalcules": rotifers, hydra, protists, and bacteria. Among those who had visited him to behold this tiny world was the astronomer Christiaan Huygens, who had turned lenses upward, his telescopes discovering the rings of Saturn and its moon Titan. While the Dutch empire he lived in has long ended, the invisible world he discovered has only grown more fascinating and complex.

Three centuries later, in 1995 in a faraway nation, Bob Williams was the director of the Space Telescope Science Institute in Baltimore, Maryland. This role gave Bob discretion to determine what the Hubble Space Telescope should spend 10 percent of its operational time doing. It had cost $2 billion to construct and blast the powerful telescope into orbit. Asking the machine to turn itself to look at something, and then to beam the data down to Earth, was time-consuming, and Hubble time was one of the most precious commodities on the planet. Williams decided to take a risk: He was going to point the telescope at an unremarkable patch of space. His colleagues tried to stop him, convinced that there was nothing there, that it was a waste of time and a waste of money. He would be ridiculed and might lose his job. "Scientific discovery requires risk," Williams said. "I was at a point in my career where I said, 'If it's that bad, I'll resign. I'll fall on my sword.' "

The telescope, sitting in its orbit above the atmosphere, turned its giant mirror to the seemingly bland area of space Bob selected. It started scanning, harvesting the faintest light sources and recording 342 pictures over 100 hours, slowly beaming them down to Earth. The single image that was gradually assembled is now called the Hubble Deep Field. The area, it turned out, was not empty. It was filled with 3,000 galaxies. Some were ancient—over 12 billion years old—and some were stranger than any seen before. A "cosmic zoo" of galaxies: elliptical ones, spirals, others with "whirling arms, fuzzy haloes, and bright central bulges." It raised the number of estimated galaxies in the universe by a factor of 5 and banished the idea that there was any such thing as an unremarkable patch of space.

The Hubble Deep Field, full of galaxies.

Bob hadn't known there was anything there, but he felt he had to look. He had a tool built for looking, and he decided to point it at somewhere new. Like van Leeuwenhoek's "animalcules," the galaxies Bob's image discovered had always been there, but they had not existed for *us* until this exact moment. I love this story. What galaxies of animal behavior could we discover if we turned our most expensive and prized tools to look upon the unexplored, overlooked living world? Sometimes it just takes someone with the guts and inclination to ignore what everyone else says and think, "Screw it, let's give it a shot."

Three years into this journey, I had a meeting with two bold and unusual men, newcomers to biology. Both were in their thirties: Aza Raskin, darkly bearded, with an expression alternating between wonder and concern; Britt Selvitelle, with curly brown hair, the founder of a Silicon Valley megalith who seemed more like a friendly volunteer on an organic farm. Aza's father, Jef, had been one of the brains behind the Apple Macintosh. Aza had advanced the family fascination with human-computer interfaces, most notably as one of the architects of Firefox, the open-source web browser, where one of his inventions was the infinite scroll function that keeps you plumbing the endless depths of your news and social media feeds. Britt was a computer scientist and engineer on the founding team at Twitter.

While each had achieved considerable success in their careers, they had grown increasingly appalled by the harms to society driven by the "attention extraction economy" they had been part of. Aza told me he had dedicated "a significant portion of [his] life force" attempting to correct these problems, cofounding a nonprofit called the Center for Humane Technology, working with governments on policy reform, and drawing attention to the issue in the Emmy-winning documentary *The Social Dilemma*. Much of this, as he knew, was damage limitation.

Technologists and nature lovers, Britt and Aza had puzzled on how they might use AI for good. What, they wondered, if we used machine learning to investigate animal communications? If these could be decoded, might people feel closer to the species we are so rapidly eliminating? And, as you do when you are a young master of the universe in a culture where disruption and innovation are king and where, if you fail, you start again and aim bigger, they decided not to enroll as students in biology to ponder this question, but formed a nonprofit start-up to begin doing it instead.

They had thrown themselves into the research, interviewing scientists and linguists working at the bleeding edge of animal communications, and engineers working with the latest pattern recognition technology. They traveled to the jungles of Central Africa to see how wild elephants interacted, and in doing so, they experienced firsthand the difficulties field biologists faced. It was during this phase of their journey that I first met them. While I loved their vision, I wondered if perhaps they were just building an exciting adventure for themselves, after spending years behind computers—until they reached out to me a few months later to say they had a plan.

The global mecca for whale-watching, Monterey Bay, is a short drive from the epicenter of the Information Age, San Francisco and Silicon Valley. In the summer of 2018, three years after my first fateful trip there, I was working just down the road from Britt and Aza. They drove down to our house, where my crew and I were staying during a film shoot. I had also invited Dr. John Ryan, a soft-spoken scientist in his fifties from the Monterey Bay Aquarium Research Institute with a penchant for skateboarding and roller coasters. John was already a believer in the power of AI to help explore whale sounds. Monterey Bay is cold; a feeding ground. It was thought that most humpback song happened far away, in their tropical breeding grounds. But John had a deep-sea listening station hooked up to his office, and he decided to trawl the recordings. It took hundreds of hours. To his astonishment he discovered the songs of hundreds of animals. John and his colleagues had then trained AIs, which made short work of six years of recordings, adding Blue and Fin whales to their listening skillset. They uncovered humpbacks singing in Monterey across nine months of the year. They learned the cold waters sometimes rang with whalesong for over twenty hours a day. John's recordings covered the time Prime Suspect had been in Monterey. He told me that he bet our whale's voice had been captured somewhere on his tapes. Sitting amid our stacked life vests, camera gyroscopes, charging batteries, and whirring hard drives, we ate fajitas and listened intently as Aza and Britt explained the plan they'd hatched. They were going to take the incredible computational power of the tech behind Google Translate and apply it to decoding animal communications.

To understand what the hell this meant, Britt and Aza had to give John and me a lesson in how AI had revolutionized translation. People had been using computers to translate and analyze language for decades; the field is known as natural language processing. Until recently, however, you had to laboriously teach the machine how to turn one human language into another. Computer programs were given decision trees to work through when faced with text in one language, and had to be instructed on what to do in every situation; they needed bilingual dictionaries and to be told the rules of grammar and so on. Writing these programs was time-consuming, and the results were rigid. Situations the programmers hadn't anticipated would arise and crash the program, such as the computers' inability to overcome misspellings.

But then came two developments: The first was the blossoming of new AI tools, like artificial neural networks—the same computer programs based on structures in the human brain that Julie used to discover unique dolphin whistles. Especially powerful in this regard were neural networks arranged in multiple layers called deep neural networks (DNNs). The second development was that the internet had made enormous volumes of translated text data freely available—Wikipedia, film subtitles, the minutes of meetings of the EU and the UN, millions of documents carefully translated into many languages.

These texts were ideal fodder for DNNs. Engineers could feed the algorithms both halves of the translation and ask the DNN to translate between them, *but without using any existing language rules.* Instead, the DNNs could create their own. They could try lots of different ways of seeing how to get from one language to a correct translation in another, and they could gamble with probabilities, again and again. They could learn the patterns for how to correctly translate. When it worked, the DNN would remember and test if it would work in a different context. The machines were learning in much the same way Jinmo Park's computer vision algorithm learned to match whale tail flukes for Happywhale. Jinmo didn't need to teach his program what a whale was, or how humans match one tail fluke to another. He simply needed lots of labeled examples and enough further unlabeled data for his algorithms to run through over and over until it found a way to make the patterns match.

While the first language-translation machines that used DNNs were decent, they were still nowhere near human competency. Most crucial, they still required our supervision: We had to give them examples of translation for them to work. Then came a very left-field development. In 2013, Tomas Mikolov, a computer scientist at Google, and his colleagues showed how, if you fed lots of texts to a different kind of neural network, you could ask it to look for patterns in the relationships between the words within a language. Similar or associated words would be placed close to each other, and dissimilar and less associated ones farther away. Aza quoted the linguist J. R. Firth: "You shall know a word by the company it keeps!"

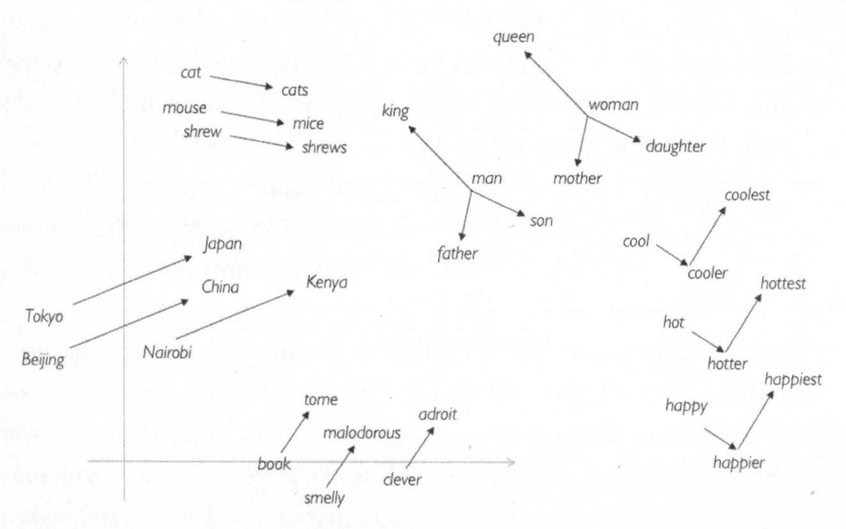

An example of word relationships in English.

For example, he explained: "ice" occurs often next to "cold," but rarely does "ice" occur next to "chair." This gives the computer a hint that "ice" and "cold" are semantically related in a way that "ice" and "chair" are not. Using written language to find these patterns of association, the neural network could embed each word into a map of the relationships of all the words in a language. I pictured this as a sort of star chart where each star was a word and each constellation within the galaxy of the language represented how those words were used relative to one another. It's actually impossible to visualize these "galaxies," as the numbers of words and their myriad geometric relationships mean they have hundreds of dimensions. But here is Britt and Aza's example of the top ten thousand most spoken words in English compressed down to a 3D picture.

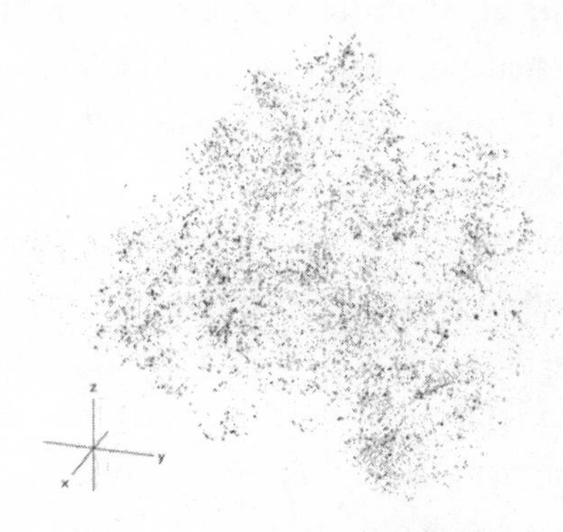

Each dot is one of the top ten thousand spoken words in English arranged as a galaxy of relationships.

What Mikolov and his colleagues discovered next was mind-blowing: you could do algebra on language! Britt and Aza broke it down: If you asked the program to take "king" and subtract "man" and add "woman," the closest word in the cloud, the answer it produced was "queen." It had not been taught what king or queen was, but it "knew" that a woman king was a queen. Even without knowing what a language meant, you could make a map of it and then explore it mathematically.

I was stunned. I'd always considered words and language as emotional, fuzzy, changeable things—and yet here was projected English, automatically assembled by a machine given billions of examples, into patterns of the relationships between words that we unthinkingly carry in our own heads, the harvest of our own neural networks from the big data of our own lives: the books, conversations, movies, and other information our brains have been fed and unconsciously tucked away.

This discovery was useful for finding relationships *within* a language, but what did it have to do with translation? This is the really neat part. In 2017 came a game-changing realization, one that had convinced Britt and Aza that these techniques could help with animal communications. A young researcher named Mikel Artetxe at the University of the Basque Country discovered that he could ask an AI to turn the word galaxies of different languages around, superimposing one onto another. And eventually, as if manipulating an absurdly complex game of Tetris, their shapes would match, the constellations of words would align, and if you looked at the same place in the German word galaxy where "king" sits in the English one, you would find "König."

No examples of translation or other knowledge about either language was required for this to work. This was automatic translation with no dictionary or human input. As Britt and Aza put it, "Imagine being given two entirely unknown languages and that just by analyzing each for long enough you could discover how to translate between them." It was a transformation of natural language processing.

And then came other new tools, too. Unsupervised learning techniques that worked on audio, in recordings of *raw human speech*, automatically identified what sounds were meaningful units—words. Other tools could look at word units and infer from their relationships how they were constructed into phrases and sentences—syntax. These were computer programs inspired by the circuitry of our brains, finding and linking patterns in our languages, which is how modern translation machines, like Google Translate, work today. And they work incredibly well, capable of translating sentences from English into Mandarin or Urdu, instantly and with reasonable accuracy. But how would they discover patterns in the communications of other animals?

For decades, humans have been trying to decode animal communication systems by looking for a Rosetta Stone—some sort of key to unlock them, a way into the unknown. Working with the smallest units, the simplest or most obvious vocalizations—like alarm calls and signature whistles—we attempted to identify a signal that might be meaningful to an animal, and then try to link it with a behavior to decode it. There was no other way, because we had no idea what any of the other sounds the animals were making meant—or if they had any meaning at all. Yet here was this new computer tool, *unsupervised machine translation*, that thrived despite not being instructed what any of the human languages it was given to translate meant. Britt and Aza didn't need an automatic translation machine to interpret my facial expression when they told me this: holy crap. So would this work with animals? I asked them. Could you approach investigating animal "languages" by mapping *all* the vocalizations a species makes into a galaxy and comparing the patterns in these to the patterns in other species? Yes, they said. That was the plan.

My mind raced. If I understood this correctly, we could map animal communication systems as we have never been able to before. We could start exploring them in-depth by comparing them to one another. We could watch those communication galaxies change and evolve over time. We could inch outward from communication systems that share likely similarities, to those less similar. From comparing different families of fish-eating killer whales, to marine-mammal-eating killer whales, to pilot whales, to bottlenose dolphins, to blue whales, to elephants, to African grey parrots, gibbons, and humans. If—and it's a big if—our automatic human-language-analysis tools worked at finding patterns in other species' communication systems, they could help us construct a context for *all* animal communications. It could give us an idea of the diversity and number of galaxies in the communication universe and where we humans sit within it.

Of course, the vocalizations of whales, dolphins, and other nonhumans might just be emotional noise, bereft of meaning, deep structure, or syntax. In which case, perhaps feeding lots of their communications into these algorithms would be like asking a facial recognition app to scan a pizza. But after all I'd learned, this felt unlikely. And even if cetaceans did have something like natural language, these techniques might still fail for other reasons.

One theory that explains why machine translations of human natural languages works so well is that all our languages are fundamentally capturing the same information. People living in Mongolia and Uganda live similar lives, in the sense that they perceive similar worlds, filled with similar objects and agents, with similar relationships, all bound by similar physics. Because the same things are therefore possible in these distant human worlds, their languages have ended up with a similar relational structure, allowing us to translate Swahili into Mongolian.

Whales and dolphins experience very different worlds to us, and *if* they have a world model captured in language, it is also likely to be very different. There may well be no similarities in the patterns of relationships between the units of humpback whale-speak and those of English, but knowing this would still be illuminating. Discovering rich, complex structures and relationships within nonhuman communication systems that bear no resemblance to those in human language would be a revelation in itself, hinting at parallel animal worldviews that we could explore. It's language, Jim, but not as we know it.

For Britt and Aza, modern machine learning is a "fundamentally new tool," for discerning patterns both within and between languages. A tool that could, in Aza's words, allow us to "take off our human glasses." I thought of Bob Williams and the Hubble telescope. Surely this, too, was worth a shot.

Throughout the dinner, as Britt and Aza described their plan, John Ryan listened attentively. What Britt and Aza now needed, they explained, was data to feed the algorithms. John had brought with him a hard drive containing thousands of hours of humpback whale vocalizations. And so an unassuming box changed hands, the fruits of years of recordings by another box deep below the sea. A box full of whale voices to be plugged into a box full of intelligences, to see what patterns they could find in the mysteries inside.

Britt and Aza called their nonprofit the Earth Species Project (ESP). I kept in touch with them over the following years. We spoke online, as wildfires raged around their homes, and through the COVID-19 pandemic as our hair and beards grew vast. They formed dozens of partnerships. In Alaska, with humpback researcher Michelle Fournet;

in Canada, with Valeria Vergara, who studies mother-calf communication in beluga whales. Diana Reiss and Laela Sayigh offered them thousands of hours of recordings of dolphins. And that's just the cetaceans. There were elephant scientists, databases of fruit bats, giant otters, zebra finches, macaques, and more. Cornell University began to share their vast animal acoustics collections, as did Oxford University. They were working with SETI (the Search for Extraterrestrial Intelligence), looking for overlaps in searching for language in the sea and in space, and investigating human whistle languages, such as Silbo Gomero (the whistle language of the Spanish island of La Gomera). They theorized that if they can find ways to train computers to translate between human whistles that have meaning, these tools could perhaps analyze dolphin whistles. And by forging partnerships between researchers across a variety of species, they thought that what they learned from one species could be turned into tools applied to others.

There is a long history in computer programming of sharing your programs and their code, of giving them away—as well as your data sets—so anyone else can see them, learn from them, and improve on them. This is known as *open source*. Aza related a saying from the open-source world: "No matter where you work, most of the smartest people work elsewhere." While the open-source movement has made some inroads into traditional academia, many biologists and their institutions are still reluctant to share their hard-won data, to give away their tools and inventions, and journals still charge vast sums to access their publications.

These, for Aza and Britt, were bottlenecks, throttles on discovery. So they modeled their biology enterprise on what had worked well in computer science. Existing recordings were cleaned up and relabeled and put online in the ESP library—an open-access repository of animal communications for *anyone* to explore. Thus, people far from the sea, humans more familiar with crafting computer games or consumer-tracking software—who had never dreamed they might ever see a whale—could be enlisted in the race to speak whale, too.

In late 2021, they reached out to me with great excitement. Going through their treasure troves of recordings, they had encountered the same challenge as Jack Fearey had with his recordings of dolphin stampedes: the "Cocktail Party Problem." If you are trying to decode a conversation, or even what one speaker in a conversation is saying, it is impossible if lots of people are speaking all at once. This problem is even harder in the sea, where sound bounces around all over the place. And cetaceans don't open their mouths or give any outward sign when they vocalize like we do. Pinpointing which dolphin said what was like trying to figure out who'd called your name at a ventriloquists' convention. When scientists couldn't identify which individual animal was speaking on their recordings, they often couldn't use them, and some of the most interesting animal "conversational data" went to waste. As ESP put it: "We can't decode language unless we can detangle conversation."

ESP now had six full-time AI experts and a multimillion-dollar budget. They built a pattern-finding tool trained on the overlapping chatterings of fruit bats, bottlenose dolphins, and monkeys. This could tease out individual animal voices within seas of sound—a cocktail party dampener that could offer the first stage of unlocking "a whole new world of social communication data" and could be applied to any animals that vocalized. The code was up on their open-source repository before their findings were published.

Britt compared the proliferation of computer tools developed from machine learning to the Cambrian explosion—the juncture about 540 million years ago where a great variety of complex life-forms suddenly emerged. It is startling to me to think of computer programs through the framework of evolutionary biology. But if one way of looking at the history of life on Earth is a story of the construction of ever more complex living systems and the diversification of information-exchanging beings, then perhaps the comparison was apt.

Despite the proliferation of tools for the untangling and pattern hunting of vocalizations, however, learning to speak whale still faced an almighty problem. To understand what these vocalizations might mean, we really had to see how they related to the behaviors of the whales and dolphins doing the speaking. Yet the wild lives of cetaceans were still, for the most part, a mystery.

* * *

In the sea there are no trails, trees, or waterholes to hide out by. To study cetaceans, you need to go and find them. You must drive a seaworthy vessel to within a few hundred yards of your subjects, who can be extremely elusive until they surface to breathe. They move a lot, some traveling a hundred miles in a day. Then there's the ocean: deep and dark and capricious. Ports, boats, and captains are limited. Salt and sun break sensitive tools. It is hard to see cetaceans through the surface glare of the ocean, and sometimes it's just as hard to see them even if you're in the water. People get seasick, and funding is tight. When the weather gets bad, biologists return to shore, and many research vessels don't work at night. There are some species we still only know from glimpses and dead specimens alone. In short, cetaceans are some of the hardest animals to record. But recently it has become possible to take a different approach: Biologists have been recording from onboard the bodies of the whales themselves.

In the summer of 2018, a month or so after I first met Britt and Aza, I tagged along on a massive research project taking place for months along the shores of California with a whale biologist, Professor Ari Friedlaender, who'd been described by several awestruck young scientists I met as a "rock star"—a description that seems to appall Ari himself. Perpetually sandaled, bearded, and with long, flowing hair, Ari is the closest marine mammal science has to "the Dude" from *The Big Lebowski*. He is one of the pioneers of whale tagging. A whale tag is a little box containing miniaturized sensors such as cameras, microphones, accelerometers, and thermometers, most of which you have in your pocket, on your phone—but these are packaged within streamlined and rugged waterproof housings. Whales constantly replace and shed their skin (hundreds of times faster than humans do), so the tags are attached with suction cups. That is Ari's job. And he invited me along to watch him at work.

Once again, I found myself in Moss Landing at dawn, where Charlotte and I had paddled that extraordinary morning three years earlier, heading out to find whales. Three larger boats worked as motherships, full of graduate students with binoculars scanning the horizon for whales while Ari and the tagging teams zoomed around in three smaller, more agile rigid-hull inflatable boats (RHIBs). These bouncy low craft with no keel allow you to

Ari Friedlaender in action off the Antarctic Peninsula, deploying a Customized Animal Tracking Solutions (CATS) tag on a humpback whale (research conducted under NMFS, ACA, and IACUC permits).

maneuver without banging into whales. At the sight of their quarry, a blue whale, radios chattered and Ari's RHIB peeled away toward it.

There were three drone teams, their various aircraft guiding Ari's tagging raft. Looking over their shoulders as they helped to guide Ari, I was able to get some idea of the awe-inspiring scale of a blue whale, the largest animal to have ever lived, as the Lilliputians in their boat closed in on the cruising marine Gulliver. Like so much of our modern technology, drones were pioneered by the military and soon after turned up in children's Christmas stockings worldwide, blessing us with astonishing perspectives while messing with our privacy. For cetacean biologists, however, drones have proven to be a game changer, flying kilometers at the flick of a thumb. They can see deep through the water from above and spot whales that would be invisible if you're looking from a boat. They can record social interactions across whale pods from on high, or hover right above the surface to gather a sample of whale snot when one exhales. Unlike helicopters, they cost little, don't seem to disturb the whales, and no one is likely to die if they crash.

The drone above Ari measured the whale, which was swimming just beneath the surface, taking photos of its body (shape, fat layers, noticeable scars, and other identifying features), and then radioed the team when the whale was about to breathe. Ari himself was at the front, where a raised metal frame had been erected. When the whale surfaced to exhale, the RHIB drove alongside it,

Ari's legs fixed against the metal bracing, his hands clasping an eighteen-foot carbon fiber pole, at the end of which was the tag. As the whale's gray-blue back slipped above water and its hydrants of nostrils gushed air, he leaned far over and whipped down the pole with exquisite timing, plonking the little tag onto the passing whale—a move so artful and practiced it would have made Melville's harpooner Queequeg proud. The sucker held and was borne away, recording everything. As the tag was being deployed, the boat's driver fired a crossbow with a special empty-capped bolt at the whale, removing a tube of skin and blubber as it bounced off the animal. This didn't seem to bother the whale much. Once retrieved, a DNA sample would be taken from this pellet, telling the scientists further secrets of the whale—who it was, where it had traveled from, what it ate, who it was related to, its health, its sex. These people in tiny rafts, one lancing a probe, the other firing a crossbow into the backs of whales, looked so like the pictures of the early whalers, it occurred to me that Ari might well have been a whaler if he had been born a hundred years ago. He clearly loved these animals, but no other excuse for such adventure near giant sea beasts existed back then.

A CATS motion-sensing and video tag on a humpback whale in Monterey Bay, looking up along its surprisingly flaky skin as a California sea lion swims past its nose (research conducted under NMFS and IACUC permit).

A few hours later, the radio receivers started beeping on the mothership to signal that the tag had fallen off. The beeps guided the scientists to the floating tag and the data inside. Video from the tag deployed that day revealed another blue whale we hadn't seen, swimming alongside it. And a world first: It measured the whale's heartbeat, taking an electrocardiogram of a heart the size of a small car, discovering a pulse that ranged from two beats per minute (*two beats per minute!*) to thirty-seven.

Once we returned to dry land later that afternoon, I went with Ari to his office at the University of California, Santa Cruz. It was a beautiful evening; orange light cast long shadows from the whale skeleton erected along the side of the laboratory buildings. Ari showed me the fruits of his years bugging whales, including the flight paths of humpbacks hunting using "bubble nets." In these tightly coordinated behaviors, two to four humpbacks team up to trap whole schools of fish, exhaling walls of air around them and herding them from the deep. Ari's tags revealed these extraordinary underwater acrobatics, showing the rising spirals the whales swam around the fish. The bubble helixes trapped the fish into tight balls against the surface, at which point the whole team of whales coordinated to simultaneously gulp their prey by the thousand.

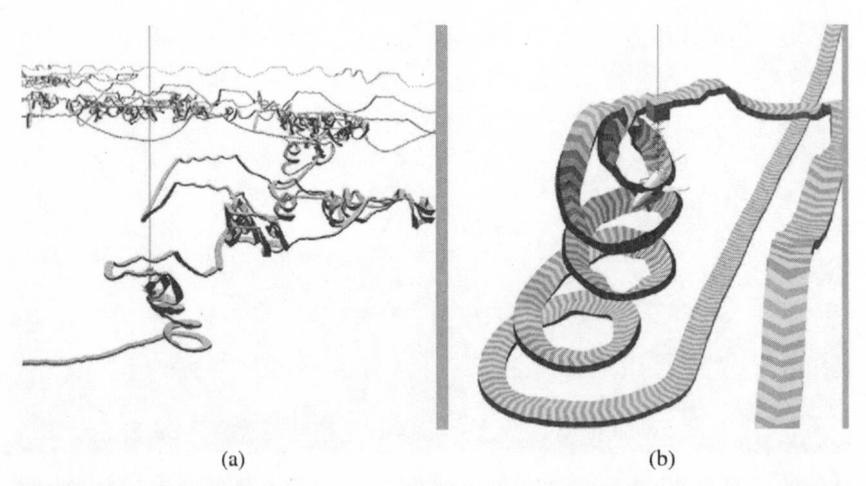

The flight path of a tagged humpback; the spiral is the whale bubble-net feeding (from David Wiley et al., 2019).

Ari has studied whales for decades. I asked him what new knowledge the tags had revealed. "We simply have no idea about the lives of these animals," he said. Almost every video showed him something new, and many confounded received wisdom. He would sit and watch and listen to every minute of every two- to five-hour tag in a single session, losing awareness of himself, hypnotized by his view of Antarctic seas sliding by and the sounds of orcas chattering. Their behavior, he admitted, was still somewhat disturbed by the boats' approach and the slap of the tag as it was attached. At least it was an improvement on early whale scientists, who could think of no other way to track whales than to shoot iron darts into their bodies and record the location, offering a bounty to any whaler who killed a tagged whale to return the tag with details of where they found it.

Three years after I met Ari, I learned he'd teamed up with Britt and Aza's ESP, who were working to apply their pattern-finding tools to his onboard recordings of the lives of humpback and minke whales, orcas, and common dolphins. Humpbacks carry out exquisite and coordinated team movements, and while they do so, they make social sounds. ESP was training machines to find patterns in the whales' movements, and then planned to combine these with machines that find patterns in their vocalizations. Linking what the whales say to how they behave, and when, is gold dust if you want to start figuring out what whale-speak might mean. There were other tantalizing prospects, too. Humpback whales are particularly agile for large cetaceans, maneuvering in close formation and sometimes touching one another underwater. Could their communications include body language and tactile elements, as well as sounds? Though these are early days, in Ari's words "it's growing into a really awesome partnership," one that lets his work "cross disciplines and boundaries and really open up opportunities for discovery."

Since Britt and Aza left our house in Monterey Bay in 2018, they have changed from dreamers to doers, surprising many in the process—especially me. What was perhaps even more heartening than their own journey, though, was to know they weren't alone on it. In the Bahamas, for instance, Denise Herzing of the Wild Dolphin Project has been developing a computer system that a diver can wear. When dolphins vocalize in the diver's presence, the computer will, in real time, translate for the diver which dolphin is "speaking," who they are speaking to, and if they are asking for one of the objects in the diver's possession for which they have learned whistles. Gift exchange is a common part of human first contact, and the diver will then be able to gift the requested object to them. Another project involves the extraordinary partnership of Diana Reiss, the musician Peter Gabriel, vice president at Google and so-called "father of the internet" Vint Cerf, and MIT professor Neil Gershenfeld. Together they've formed a think tank called the Interspecies Internet, a project working to link non-human species using AI and machine language to "transduce signals" from one species to another. Meanwhile in the chilly waters of Skjervoy, Norway, an interdisciplinary Swiss team is testing a prototype "interaction device" allowing them to mimic the sounds of humpback and killer whales, analyzing the whales' vocal responses to the humans' voices in real time. "It's promising, very promising," neuroinformatics scientist Dr. Jörg Rychen told me.

Yet the projects I had seen so far had a problem in common: They were all working with fragments. Trying to figure out what a whale might possibly be saying, when you had only a few minutes or hours of recordings, when you could rarely figure out who was speaking and what they were doing, was like trying to decode parts of a script shredded into small pieces with the characters' names blacked out. So much of this work was making the best of existing data sets and fleeting encounters. To give real grist to the machine-learning mills, we needed big data. *Big whale data*. But how could we get it?

Enter, stage right, our old friend Dr. Roger Payne.

What if you could start from scratch and design a mission to record a data set of whale communications perfectly optimized for the latest machine-learning and language-processing tools to scan? A set of recordings of multiple orders of a magnitude bigger than anything previously captured. What if you could capture not just whole *conversations* but hundreds of thousands of them, from scores of different whales totaling millions, perhaps billions, of vocalization units? Would you then have a chance at speaking whale? This is the plan of the Cetacean Translation Initiative, or CETI. On Christmas Eve, 2021, Roger Payne called to tell me that its work had already begun.

CETI is an enormous beast, comprising an interdisciplinary A-Team of badass scientists: marine robotics specialists, cetacean biologists, AI wizards, linguistics and cryptography experts, and data specialists. They were all brought together at a meeting of academics at Harvard in 2019, which was chaired by David Gruber. Gruber is a marine biologist and inventor, crafting cameras that can capture the glow of sea turtles and soft robot graspers to gently handle fragile deep-sea animals. He resembles a more dashing Egon Spengler from *Ghostbusters*. Roger is the principal advisor in whale biology. Their team is huge: scholars from the universities of Imperial College, MIT, Lugano, Berkeley, Haifa, Carleton, Aarhus, and Harvard, with help from Twitter and Google Research and grants from the TED Audacious fund, the National Geographic Society, and Amazon Web Services. Their goal, he told me, was "to learn how to communicate with a whale well enough to exchange ideas and experiences." And they are not wasting time.

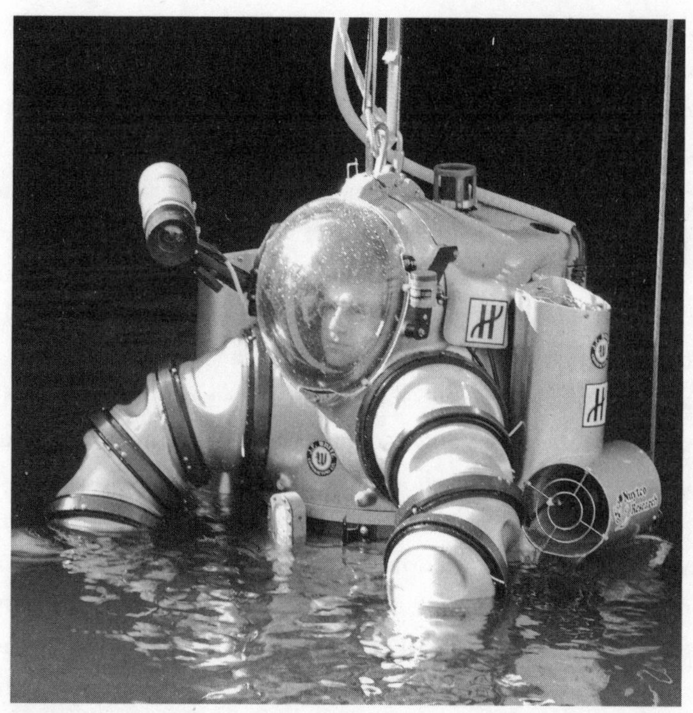

David Gruber being lowered into the sea in an Exosuit one-person submarine.

CETI's audacious plan is to throw everything they've got at the population of sperm whales off the island of Dominica in the Caribbean—a laser focus that will allow them to go, in David's words, "from small data to big data." This population is already well known through the work of cetacean biologist Shane Gero, who has identified hundreds of sperm whale individuals and their vocalizations. If you've ever been given an old box of family photographs and tried to find meaning in them, you'll understand how valuable it is to be familiar with the relationships and chronologies of the people whose lives they capture. Gero's decades of careful listening and identifying will provide a vital context for what they record—and they are going to record a great deal.

Roger told me that he has worked for sixty years while dreaming of a chance like this. When he sketched out its scope, it took my breath away. CETI will rig the seafloor with multiple listening stations, trailing others on lines up to the surface. There will be floating stations dangling arrays of further listening devices every three hundred feet down to more than a thousand yards. These will cover 7.5 square miles (twenty square kilometers) and will form the "Core Whale Listening Station," recording twenty-four hours per day as the whales go about their lives. Drones equipped with hydrophones will fly in formation over groups of active whales and surround them, carefully cutting their motors and lowering their hydrophones once in position. When the group of whales moves on, the drones will take flight once more and repeat the process. And swimming with them will be "soft robotic fish" equipped with audio and video recording equipment, able to move among the whales without disturbing them. New, cutting-edge tags with redesigned shells and an attachment mechanism modeled on octopus tentacles will be employed, allowing recording for days and perhaps even weeks, rather than just a few hours. Importantly, they will remain fixed to the whales when the creatures dive deep, capturing their vocalizations and even their viewpoints in the near pitch-black. The whales will be living in an auditory panopticon, and all of the codas and other sounds that make up their Morse code–like communications will be captured for analysis.

Sperm whale pods have around fifteen members, with each pod having its own distinct way of speaking. To represent this diversity of sound, CETI

hopes to place tags on mothers, grandmothers, teenagers, and great bull males from many different pods. There will be weather sensors and other contextual data, and they will link vocalizations to behavior and what they know of each individual whale: Was it hungry, fishing, pregnant, or mating? Was it speaking to its mother or a rival? Was there a storm? Was there much squid to eat? Were predators threatening the whales? With all the information they'll collect, the researchers will be able to track each individual whale over time, forming a "social network," sketching the stories of their lives to link their vocalizations to. In total, the recordings will form not just the largest and most complete sperm whale data set, but probably "the largest animal behavioural data set" ever gathered. For any nonhuman species.

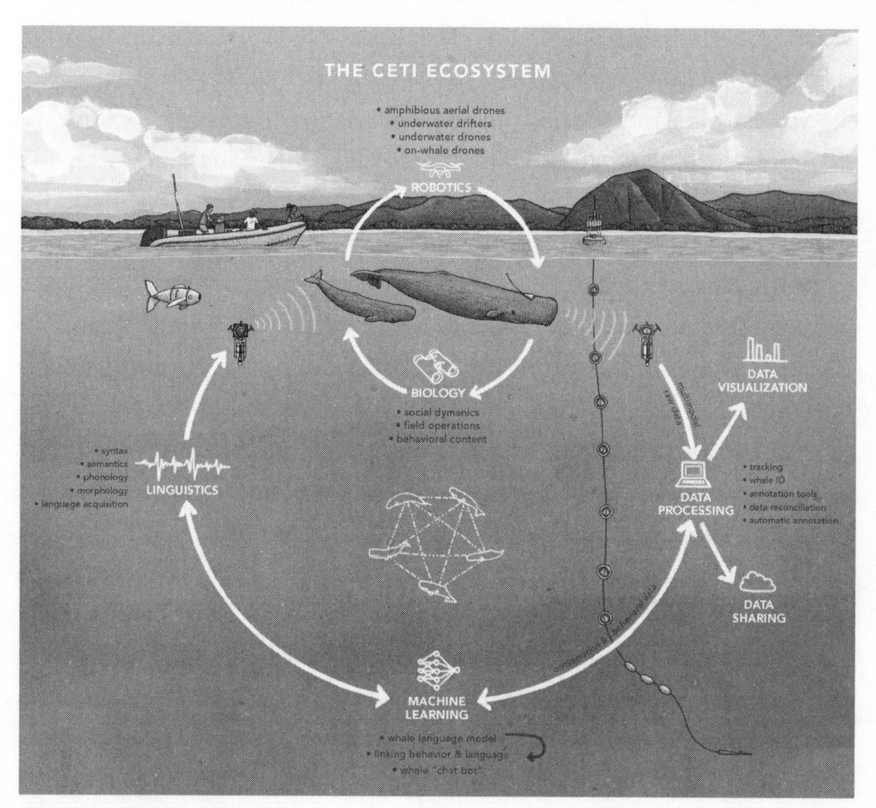

How CETI will work.

These data will need a home. Just as nineteenth-century naturalists sent home crates of specimens—preserved fish and dead bugs, stuffed birds, and plaster casts of tiger footprints, all to be displayed and studied in glass cabinets at museums—so, too, will this biological information harvest be physically housed in humming, climate-controlled data centers. The data will need to be stored and processed, with "automated machine learning pipelines" taking care of the annotation normally done by hand, but which no human hand could undertake on this scale. All of these data will be available for the open-source community, so everyone can get stuck into "the deep wonder of our attempt of meaningful dialogue with a nonhuman species."

Then the AIs will really be unleashed. They will identify where there are sperm whale vocalizations in the recordings, then separate the clicks of echolocation from those that contain the coda click patterns the whales use to communicate. They will analyze these codas, distinguishing between those of different clans and individuals. They will analyze the structures within these codas, seeking the building blocks of the communication system. This repertoire will be mapped and analyzed for relationships between the acoustic units, seeking and constructing composition rules and grammars, and higher-level structures of syntax across the codas. In revolutionary detail, the AIs will chart the sperm whale communicative galaxy.

By listening to baby whales learn to speak, the machines and the humans guiding them will themselves learn to speak whale. The team will examine not just how a whale says something, but how the discourse structure works—how the whales use their communication. Do they take turns or overlap? Do they echo one another's utterances? And they will link the sounds the whales were making to what they were doing at the time, discerning which whale spoke and who responded—and what both parties did next.

And that's not the end of it. All of the machine-learning tools will be searching for patterns to help "constrain hypothesis space" (narrow down their theories) as linguists and other team members take the patterns they've discovered and attempt to build a working model of the sperm whale communication system. To test this system, they will build sperm whale "chatbots." To gauge if their language models are correct, researchers will test whether they can correctly predict what a whale might say next, based on their knowledge of who the whale is, its conversation history, and its behaviors. Researchers will then test these with playback experiments to see whether the whales respond as the scientists expect when played whale-speak.

Finally, they will "attempt bidirectional communication"—they will try to speak, back and forth, with the whales. What did they expect to say? I asked David. "The important thing to me," he said, "is to exhibit that we care and we are listening. To show the other beautiful life-forms that we see them."

CETI is aiming to do all of this by 2026. While this may sound like science fiction, it has already begun. At the time of writing, the team is returning to Dominica with the latest whale tags, each packing three hydrophones pointing in different directions. The Core Whale Listening Station and its twenty-eight hydrophones is being assembled in the home range of the twenty-five resident families of sperm whales. The team is now moving on to testing drones that can deploy tags on whales, and the soft robotic fish that will swim among them. *If all goes to plan, by the time this book is in your hands, all this will have come to pass.* All the whales watched over by machines of loving grace.

MIT scientists testing a SoFi (soft robotic fish).

During my conversations with the CETI team, I couldn't help but think back to something Aza had said: "It's as if these AI tools are the invention of the telescope and the new data sets are the night sky. You found us just at the moment of turning the telescope up. Imagine what we'll discover." If CETI isn't an animal-pattern-finding Hubble, I don't know what is. It was almost bewildering to me, all the pioneering techniques and technologies I'd witnessed on this journey coming of age and combining and crystalizing. Passive acoustics, tags, drones, autonomous vehicles, machine learning, and natural language processing—all open-sourced and shared. This was to be a truly interdisciplinary collaboration, bringing a "data centric paradigm shift to the study of animals," as one scientist put it. Whether or not CETI succeeds, the technologies of Silicon Valley, its money and ambition, have entered this game and forever changed it.

These developments are only possible because of the last fifty years of hard, vital human research to explore and document the complex lives of cetaceans. A half century that has seen sperm whales go from being considered mute to being recognized as some of the most sophisticated communicators on earth, a period that has seen terrible destruction to whales and our oceans, as well as a rise of human movements to protect them and the seas. Roger, who is now eighty-seven, has witnessed all this. He told me that if CETI succeeds and we communicate with another species, it will "change our respect for the rest of life entirely—as in completely, utterly, shockingly, surprisingly, unexpectedly, fully." And it is this change that he believes could save us from destroying nature, and ourselves. He had spent the last two years of the COVID-19 pandemic apart from his beloved wife, Lisa, who was unable to leave New Zealand. These were years of catastrophic wildfires, the widespread melting of Arctic ice, irreversible damage to the Amazon rainforest. When he told me about CETI, I wondered whether it was this project that had helped sustain him during such hard times.

What could these scientists and tech experts discover that might make us reconceive of our relationship with these creatures? Is it too much of a leap to think we might someday decode the sperm whale click for "mother"? For "pain"? For "hello"? The answer is, of course, that we cannot know until we try. And it feels to me that this is a thrilling time, of people at ease with turning powerful tools toward the dark, with the conviction that this was perhaps the most important task on Earth. As Jane Goodall wrote when she was told by ESP of their plan, "Ever since I was a child, I've dreamed of understanding what animals are saying. How wonderful that may now be a real possibility."

And whales are just the start.

In the spring of 2021, I was with my friend Tristram and his seven-year-old daughter, Adi, in their garden. He had been teaching young Adi the names of insects and wildflowers, but some foxed him. He told me that this made him think of his late father, a learned naturalist. "I haven't had him looking over my shoulder, telling me what things were, and I miss it deeply. I never absorbed it all." But now in his pocket, Tristram had an AI bug and insect identifier. "I have it telling me what the second instar of this bug is, and it's made my life *so good*." I myself now know the names of the trees in my local park because of my AI-powered tree-identification app, PictureThis. Now when I hear a bird that I don't recognize, I get out my phone and use my birdsong app, Merlin (a sort of avian Shazam)! Advertised to me on Instagram is an app called Blossom, which not only identifies plants and tells you how to look after them, but uses computer vision to diagnose whether your plants are sick, overwatered, or sunburned. Another Instagram ad pushes me a picture of "my.bird.buddy"—a bird feeder that automatically takes photographs and videos of the birds that visit, and identifies them for you by their sounds and image. It claims to recognize a thousand species, which is 10 percent of all the world's birds. It then sends this information to an open-source platform allowing scientists to track migrations and species numbers. It costs $150. It wasn't developed by a lab or big pet company, but by some friends on the crowdfunding platform Kickstarter. It is a citizen-funded, mass-market, AI-powered, social-media-optimized conservation bird feeder.

These technologies mark a revolution in the way we're recognizing and learning about nature. Indeed, the author Alexander Pschera thinks that we are now being brought closer again to nature—to the movements, sights, lives, and knowledge of animals and other living things—*by* our technology, claiming, "the Animal Internet has the potential to revive the human-animal relationship." The clunky tools of the biologists that I'd met have been honed and made intuitive, and found their way into my pocket. And this is all happening so fast.

Ted Cheeseman of Happywhale emailed me recently with the news that his system now knew almost *every single living humpback whale* in the North Pacific Ocean.

"What started as a 'hey, I wonder if we can do this,'" he wrote, "is now a point of AI-assisted collaboration across oceans and decades in a way that I kind of can't believe is working."

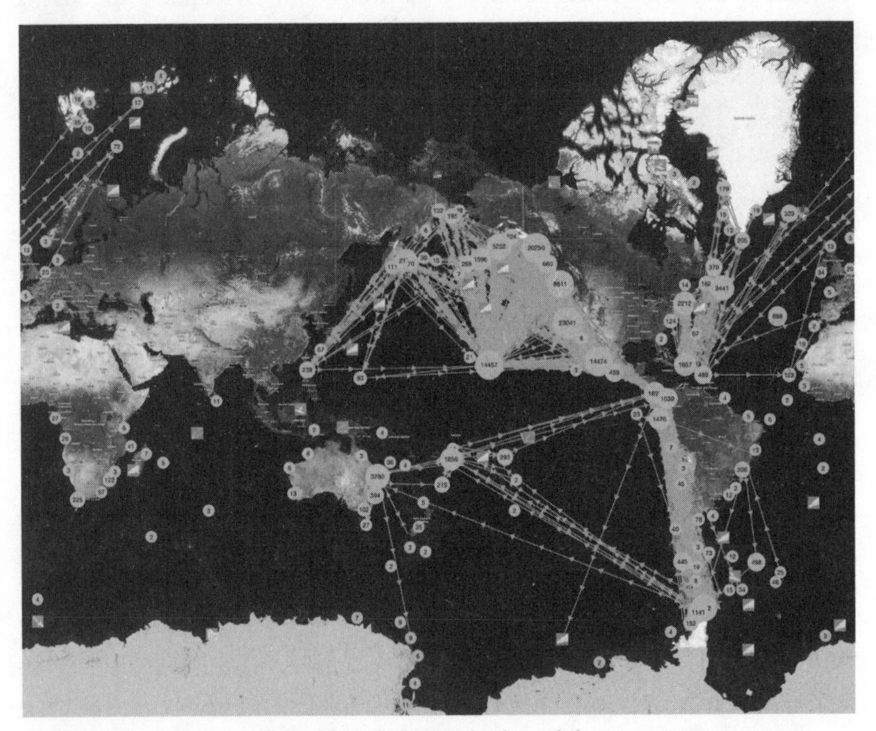

Ted's Happywhale map of all indentified humpbacks and their movements.

So, with this and all I had seen in mind—the hints of language-like abilities in animals; the extraordinary bodies and minds and behaviors of whales; the revolution in the technologies for sensing, recording, and analyzing; the international, well-funded, historically significant collaborative projects and their grand ambitions—was it inevitable that we would now discover what whales and other species are communicating, if anything? I was still unsure. I found myself, weirdly, on the fence. The scientists I'd met over the course of my investigation had all talked about learning to listen better. Yet this left me with a question I couldn't quite shake. Are we ready to listen at all?

A beluga called Hvaldimir.

11

Anthropodenial

Animals don't exist in order to teach us things, but that is what they have always done, and most of what they teach us is what we think we know about ourselves.

—*Helen Macdonald,* Vesper Flights

In 1856, in the Neander Valley near Düsseldorf, workers at a quarry unearthed a skull fragment and limbs that seemed to be from a human-like animal, with a large nose, a heavy brow, and a shorter, stockier body. They soon found more bones belonging to other individuals. These new hominids were named Neanderthals, after the valley. Since then, we have found thousands of their fossils and artifacts. We've learned that they lived from around 400,000 BCE to about forty thousand years ago, at which point they vanished. They lived across Europe, from Portugal and Wales in the west and into the Altai Mountains of Siberia in the east. Before we knew any of these things, we decided that they were inferior to us and our human ancestors. Human beings are called *Homo sapiens,* or "thinking man." One suggestion for the scientific name for the Neanderthals was *Homo stupidus.*

But we were wrong. We have since discovered that Neanderthals were not only powerful and brave, but smart, too, with finds indicating they teamed up to make weapons and traps to hunt bison and reindeer. One hundred eighty thousand years ago, these people hunted mammoths in modern-day Jersey. The stone tips for their blades and axes were mined far away and carried as blanks, ready to be carved when needed. Archaeological digs revealed that Neanderthals had jewelry and perhaps art; they made fire and complex tools and clothes; they seemed to have religious beliefs of some sort and could perform major life-saving surgery. Sixty thousand years ago, in Spain, they splattered and blew red ochre in lines along stalagmites—and over a period of ten thousand years, different bands of Neanderthals were drawn to the same cave to do the same thing. These patterns must have meant something important to them. Neanderthals were seemingly intelligent and could communicate. While we thought that humans had the biggest brains of all the primates, scans of Neanderthal skulls revealed their brains were larger than ours (although we know this isn't necessarily a perfect indicator of intelligence in itself). They buried their dead. Yes, they were different from *Homo sapiens*, but they weren't nearly as different from us as we'd expected.

Because they disappeared, the leading assumption had been that we, the dominant, superior hominid, had outcompeted our primitive relatives and killed them off. Yet this theory began to look wobbly as more information came in. Neanderthals didn't suddenly vanish in areas where our ancestors lived. Instead, we overlapped, and they gradually petered out, possibly as the animals and foods they ate disappeared in the changing climate. Despite this, they are not fully gone and in fact continue to live on in us today. Genetic studies have found that in some humans, as many as 2 percent of our genes are from Neanderthal ancestors. Our species met, mingled, and mated. When the human quarry workers found those Neanderthal bones, they were not "discovering" these other humans, but rather reencountering their relatives. Our powerful assumptions about them—that they were primitive, beneath us, and that we conquered them—made it harder for us to perceive a more complex story: that we had things in common, our lives interwove, and we could also have gained traits from one another. These "stupid" bones could have belonged to our ancestors' friends.

This initial theory about Neanderthals being inferior to us was devised by scientists and persisted until very recently, but it rested on something profoundly unscientific—namely, the deeply held beliefs of our culture.

The lenses we carry within us, which unconsciously color all we see. I'd learned so much about the challenges of understanding another animal's communications—the technical obstacles of learning to speak whale, of gathering data, finding patterns, testing observations—and so many of these had at one time seemed insurmountable. Hundreds of years ago, when Molyneux drew his map, whales were simply monsters, culprits of biblical crimes, suitable only as fearsome pictures to fill in gaps in a map of the ocean. It was likely inconceivable then that we would, a dozen lifetimes hence, photograph them from space, see the world from machines aboard their backs, decode their "names" within their calls. Yet here we are. Our technologies have certainly changed. But have our prejudices?

Anthropocentrism, or the conviction that humans are exceptional, made us group *Homo neanderthalensis* into a mental category with all the other animals, despite their clear resemblance to us. We carved a mental gulf that has taken almost a century of evidence to narrow to the point where we can see them for what they were: not "better," not "worse," but different. Yet in spite of these discoveries, to call someone a Neanderthal in the modern age is still considered an insult. Of course, there is a flip side; namely, wishful thinking, where we project onto other animals our belief that they are like us, or even superior. Bias, I now believe, is the final obstacle if we are to try to speak whale. Clearly, it runs both ways and is inside all of us, including me.

Throughout this long adventure, as I had come to understand how for many people, decoding animal communications was no longer a fantasy but a technical problem, something had been nagging at me. Although I personally loved the idea, its romance and possibility, another part of me could not accept that this might ever happen. Part of it was logical: We do not currently speak whale, whales do not speak human, ergo it cannot happen. Part of it was a biologist's skepticism born from a knowledge of our previous failed attempts to speak to other animals or teach them to speak to us. But there was something else, something deeper that recoiled *emotionally*, a belief that "speaking whale" was a ridiculous notion: that there was no point trying to speak to a whale because they can't speak; that they're not capable of having thoughts you might engage with.

How could I know this? Where did this conviction come from?

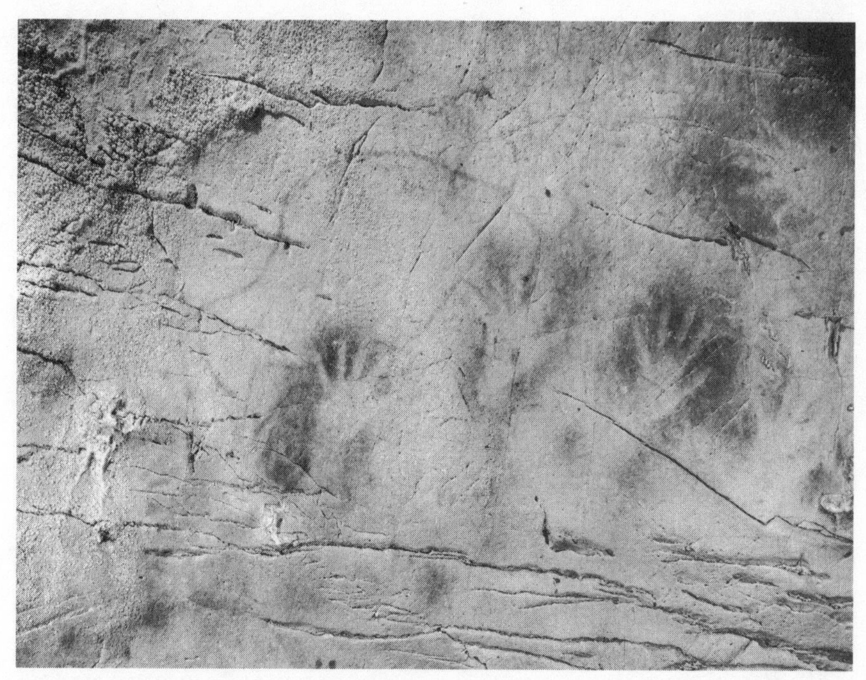

Red ochre hand stencils, El Castillo, Spain. You can see where the artist has sprayed pigment across their hands to make them. Neanderthals daubed themselves and their objects with ochre. These paintings are over 39,000 years old, so ancient some scientists believe Neanderthals created them.

In February 1649, the French philosopher and mathematician René Descartes wrote to his friend the philosopher Henry More. Descartes was one of the great thinkers of his time, and his time was one of new and radical thoughts. He believed that we could pursue knowledge by the strict application of our rational minds to the world around us. His thoughts and discoveries were a pillar of the "Age of Reason," or Enlightenment, a period in Europe where human ideas and ideals were dramatically reformed, many of which still underpin our lives and beliefs today. His most famous philosophical statement was "Cogito, ergo sum" (I think, therefore I am). This meant that he knew he existed because he could reason. For Descartes and many others, reason was uniquely human, a special gift. Through reason, we have since learned much about the world—but Descartes pulled up the drawbridge of reason behind the human species, and with this, he cut us off from considering the potential for rational universes in other species. At the time, in Western Europe, there was an explosion in the construction of machines—or *automata*—that moved by themselves. Descartes felt it was a reasonable idea that "*nature should produce its own automata much more splendid than the artificial ones. These natural automata are the animals.*" Other species were not like us, he claimed. They were simply biological machines.

A keen experimenter, Descartes, like many of his peers, observed and participated in vivisection, fascinated to see the still-pumping hearts and other workings of living animals laid out. How could the sensitive philosopher not feel this was cruel? It wasn't that he thought fish, dogs, and the other poor animals used were unable to feel pain; as he wrote, "I deny sensation to no animal, in so far as it depends on a bodily organ." For Descartes sensations alone did not matter. What was vital was the unique human gift of *rational thought*. Animals could feel, but they did not really think—and proof of this was their inability to communicate beyond biological imperatives. He wrote about a magpie who had been taught with rewards of tidbits to say "Good day" to his mistress whenever she approached him. To Descartes, while the magpie might appear to be voicing its thoughts, the explanation was far simpler. It was merely a machine that had been trained to make a sound expressing its emotional hope of eating. "Similarly, all the things which dogs, horses, and monkeys are made to do are merely expressions of their fear, their hope, or their joy; and consequently, they can do these things without any thought," he wrote.

To reason was uniquely human. And language, as the expression of reason, was proof of this. If you could speak, it meant you had important, different thoughts that other animals didn't have because they couldn't speak. So you could treat them as you would never do a fellow, rational human.

Descartes was by no means the first human to place our species above others. In Western Europe, Christianity dominated life, politically and intellectually. In this culture, man's role was that of dominion over beasts, of shepherd and civilizer, and much writing about nature involved instruction on how to cultivate and control it. This was the natural order of things.

Even before Christianity, the *scala natura*, or Great Chain of Being, was a philosophical concept first developed by Plato and Aristotle and other ancient thinkers and adapted in various forms by subsequent societies. A hierarchy of everything and everyone on Earth, with gods at the top followed by lesser supernatural beings, then kings and other elite mortals, then bog-standard people. Underneath those humans were the animals considered most important and useful, followed by animals considered less useful, and so on, with the lowest rungs composed of inanimate things, like minerals and rocks. Subscribing to a *scala natura* was a simple way of knowing your place and keeping others in theirs.

By Descartes's time, the *scala natura* was under threat. Explorers voyaged across the seas, to new continents and icy poles. They encountered people and animals that confounded their expectations. Astronomers discovered new planets and plotted the movements of the spheres. Some observations did not sit well with those in power—kings and popes whose legitimacy depended on a hierarchical world, with Earth the center of the universe and with them at the top of the earthly order. When, in 1600, Giordano Bruno suggested that in fact the universe was vast, full of other stars and planets, and that our sun was the center of our solar system, he was pronounced a heretic and burned at the stake. But while you can burn the messenger, you can't hide the sun.

Discoveries in biology have long influenced how we think of ourselves in relation to nature, and where we draw the line between ourselves and others in general. Lots of early biology involved cataloguing and organizing nature, from Aristotle's *History of Animals* through to the twelfth-century botanical texts of the Arab polymath Ibn Bājja, and the later writings of Saint Albertus. These works were few and far between, as were literate people who had the opportunity to read them. But with the experimenters of humanism, the rediscovery of classical cultures in the Renaissance, and the challenge to the Catholic Church presented by the Reformation came opportunities to apply human reason to the natural world around us. With new tools, rich patrons, and a culture of greater interest, people like Descartes endeavored to understand animals, to see how they worked and to relate this to how we worked and explain why we were special.

What they found chipped away at ideas of European and Christian and royal specialness, although they continued relying on those ideas to justify the colonizing and exploitation of many of those they "discovered."

Animals, plants, and sometimes, horrifically, even other people were shipped back from the New World to Hampton Court, the Malay Archipelago to Constantinople. Presented to royalty and exhibited in new zoological and botanical collections to the public were animals that confounded ideas of what animals should be. A kangaroo was first described shortly after the year 1500 as a "monstrous beast with the hands of a man, the tail of a monkey and that wonderful provision of nature, a bag in which to carry its young." The poor kangaroo was then captured and its body carted off to astonish the court of Ferdinand and Isabella.

Four hundred years later, zoological discoveries continued to inspire fascination and awe—a live giraffe's arrival in late nineteenth-century Paris inspired a craze, including a towering new haircut among its citizens. In England, biological discoveries were made by experimenter philosophers and physicians supported by their own funds or royal patronage. The Royal Societies and other science institutions organized themselves to share their findings and met to perform demonstrations and experiments. Merchants with equipment-making skills, such as van Leeuwenhoek, dabbled in discovery. By the nineteenth century, other kinds of amateurs joined their ranks; in the British Isles, country vicars and gentleman naturalists educated in observation and experimentation and with time on their hands noted the arrival and departure of migratory birds, the life cycles of insects, the flowering and hybridization of plants, the layering of rocks and the discoveries of fossilized giant beasts within them. They observed, recorded, predicted, interrogated, and shared the fruits of their labor. Equipment for looking and testing proliferated; inquiring humans explored the microscopic, the atomic, the chemical, the interstellar. Their findings shook everything from how old we thought the Earth was to the size of the universe, to what matter is made of.

Just as those making discoveries about gravity and the composition of the air sought to place these within wider models of physics and chemistry, so "natural philosophers," later called biologists, sought to find unifying principles for how life worked. By this point, discoveries from across the sciences—such as those of naturalists like Charles Lyell, Charles Darwin, and Alfred Wallace—had radically changed the story we told about our origins, our home, and its place in the universe. We had thought ourselves a special planet in the center of the solar system, itself the center of a small universe. Now we understood this was not so. We knew, too, that instead of being created in our God's image, we had evolved slowly and blindly from fishy sea creatures, and our closest kin were the apes. There were other galaxies, other planets, and life existed at scales we could not comprehend: whole ecosystems in drops of water, trees older than the Bible, ant societies that outnumbered even our own terraforming species. With each discovery, the scope and wonder of all life expanded, but our own place within it, our leading role, seemed diminished.

Yet beyond this scientific and intellectual progress, one story remained resilient: the story we told about other animals. Here, we were still exceptional. In the words of the contemporary philosopher Melanie Challenger, "the world is now dominated by an animal that doesn't think it is an animal."

It's worth stating that not everyone subscribed to this way of thinking. In 1580, a century before Descartes's own postulations, the French philosopher Michel de Montaigne wrote, "When I play with my cat, how do I know that she is not playing with me rather than I with her?" Such views, however, were outliers. It would be many centuries before anybody attempted to scientifically investigate questions in this vein about the animal mind. Even by the nineteenth century, most biology often consisted of learned men cutting up dead animals to figure out how they worked, and reassembling them to organize them into collections. In fact, it was only as recently as 1898 that Edward Thorndike, a psychologist, published the first-ever psychological study involving nonhuman subjects. By 1911, so little had changed that he complained that "the beasts of the field, the fowl of the air and the fish of the sea" had all been examined by hundreds of workers with "infinite pains" to figure out how their bodies worked. How about looking at their *intellect*, proposed Thorndike. Gradually, other scientists followed suit, mostly in laboratories, where experiments were devised to study the behavior of easily kept animals, like rats, pigeons, or dogs, such as those of the animal physiologist Ivan Pavlov and his famed pooches, trained to unconsciously salivate at the sound of a bell.

In this vein, ethology—the study of animal behavior—was developed later in the early twentieth century. A founder of the field, Nikolaas Tinbergen, who spent time watching wild birds, described this method of unpicking behavior as "watching and wondering." Another, Karl von Frisch, studied honeybees and discovered that returning foraging bees would dance to their hive-mates to indicate the right direction and distance to their food source. There was skepticism that such primitive creatures as honeybees could have a dance language. Today, robot bees have been built following von Frisch's dance rules, which succeed in directing other bees to new food sources. In 1973, von Frisch, Konrad Lorenz, and Tinbergen shared the Nobel Prize in Physiology or Medicine. The study of animal behavior had come of age.

In this photograph, Konrad Lorenz, one of the Nobel Prize–winning ethologists, leads a line of goslings. He discovered that the newly hatched birds would "imprint," bonding with the first moving object they saw and following them. In this case it was him.

Over the next few decades, what we would learn about animals would upturn our most ingrained assumptions about them. Biologists got to work ascertaining how an animal's behavior related to its *fitness*: its ability to survive, reproduce, and perpetuate its genes. Behavioral ecologists studied how the behavior of an animal was selected to deliver benefits within the constraints of its habitat. Cognitive psychologists looked to explain animal behavior in terms of how the animal receives, arranges, and acts on information from the world. Entire human lifetimes were spent watching the behaviors of lions, gulls, chimpanzees, elephants, crows, octopuses, and parrots. Today, biologists are slowly accepting that, as well as having complex learned behaviors, individual animals might be quite different from one another and have distinct personalities. For Jaclyn Aliperti, animal personality researcher at UC Davis, rather than looking at an animal and thinking it belongs to a particular species, "I view them more as individuals. I view them as 'Who are you? Where are you going? What are you up to?'" Since we began looking at what animals *can* do, we've found they can do a lot, with some abilities far beyond our own. In fact, here's a quick and inexhaustive list of things that other animals seem to be able to do—that we used to think were uniquely human:

Making tools
Cooperating to achieve tasks
Planning ahead
Having menopause
Understanding abstract concepts
Memorizing hundreds of words
Remembering long number sequences
Doing simple mathematics
Recognizing human faces
Making and having friends
Kissing with tongues
Experiencing mental illnesses
Grieving
Using syntax
Falling in "love"
Feeling jealous
Accurately mimicking human speech
Experiencing awe, wonder, or even "spiritual" experiences
Feeling pain

Feeling pleasure
Gossiping
Killing for "pleasure" (where no food, defense, or other reason is known)
Playing
Exhibiting morality
Demonstrating a sense of fairness
Performing altruistic behavior
Making art
Keeping time, moving to the beat, and dancing
Laughing, including when tickled
Weighing up probabilities before making decisions
Emotional contagion (feeling pain when seeing others in pain)
Rescuing and comforting one another
Displaying accents and cultural differences in sign and verbal signals
Having and transmitting cultures
Predicting the intentions of others
Intentionally intoxicating themselves with alcohol and other substances
Manipulating and deceiving others

While some of these discoveries are contentious, because our terms are so human that to extend their full meaning feels to many observers like an overreach, there is some evidence for every single one. Even among those who are aware of these findings, there is sometimes a resistance to engage with the implications of these discoveries, a readiness to dismiss the chance of them being correct. Did these comparisons of animal experiences and abilities with our own tug at something deeper, an instinct that was not perhaps that of a scientist wanting precision but of a human still searching for exceptionalism?

The primatologist Frans de Waal has a good term for when we dismiss an animal's behavior when it seems able to do something a human can do: "anthropodenial." An interesting example is mourning. While this is a powerful urge in humans, with a common suite of behaviors and characteristics we experience after loss, there are hints that it is not confined to us alone. Elephants will turn their departed fellows' bones over with their

trunks and sniff them, gently rest their feet on their jaws and skulls, probing the tusks they would have touched when reaching out to greet one another in life. Relatives seem more drawn to the remains than others. On occasion, they have been observed covering their dead with soil and vegetation. Sometimes, when they come across places where their friends died, they will pause and stand very quietly—even when the bones are no longer there.

In cetaceans, too, there are recordings of behaviors relating to what we might call grief. Killer whale and dolphin mothers have been observed pushing their dead calves around for days and sometimes weeks, like the orca mother in the Southern Resident killer whale group off British Columbia, known as J35, or Tahlequah. She captured and then broke the hearts of people following her story across the world by carrying around her dead calf for seventeen days. Researchers following Tahlequah were worried about her thin state; the other orcas in the group seemed attentive to it, too, taking turns to carry the dead calf while the "grieving" mother rested. After a thousand miles, she let her now-decomposed calf go. In fact, researchers recently discovered that "post-mortem-attentive behavior" had

A different orca mother, L72, a twenty-four-year-old, carries her dead newborn off San Juan Island.

been recorded in twenty different species of cetaceans. As the cetacean neuroscientist Lori Marino says, "There's no reason to think grief would be restricted to humans."

Of course, we can also lean in the opposite direction to anthropodenial. We can anthropomorphize, often projecting our own inner worlds and motives onto animals that do not share them. Sometimes we even go further than this, bestowing upon animals abilities that humans themselves don't possess. I learned this firsthand in Hawai'i, when I met a woman named Joan Ocean. Joan is one of the founders of Dolphinville—a loose, self-given label for the two hundred or so people who have traveled from across the world to live on the southwest coast of Kona to swim and otherwise commune with the resident spinner dolphins. Joan, a beaming and luminously tanned woman, became fascinated with cetaceans after meeting the legendary and controversial dolphin researcher John Lilly in the seventies, and had joined an early interspecies communication mission to Vancouver Island—where, she told me, a whale had spoken to her. Joan felt she had been chosen as a sort of cetacean ambassador to bring their teachings to humanity "in a manner that will ensure acceptance." She has since devoted her life to this task, swimming with dolphins for thirty-three years, spending more time in the presence of wild spinner dolphins than perhaps any human ever has. She explained to me that the dolphins had told her of distant stars, universes invisible to us humans, of plasma ships, pyramids, and the techniques of shape-shifting. Joan's belief, fundamentally, is that dolphins and other cetaceans are not of this Earth.

But how could this fit into what we have seen? I thought. What of the killer whales that seem to toy with prey species, killing them slowly and then discarding them? What of the many cetacean species that race across seas to ride in the bow waves and wakes of our boats, chattering and jostling, where our best guess is that they do so for pleasure alone? What of the bottlenose dolphins observed beating harbor porpoises to death? What about the evidence that they care for their sick and disabled? Or those strange orcas off Gibraltar that have been smashing up the rudders of sailboats, leaving so many adrift that the government banned small boats from their territory? Or those dolphins recorded "talking in their sleep," vocalizing imitations of humpback whales? Why would space creatures come here and do all this?

I like Joan and admire her lifelong devotion, and I respect this belief as hers to have. Yet I feel the same flaws underpin dolphin worship—anthropomorphism and anthropodenial. Each is too simple, lacking in evidence, and relies on the projection of exceptionalism of either humans or dolphins. The nature writer Carl Safina wrote about how humans are "simultaneously the most compassionate and the cruelest animal, the friendliest and the most destructive." We are, he wrote, "a complicated case." That is surely what makes us so interesting—and perhaps the same could be said of other animals, too.

Rather than overestimating the abilities of animals based on our personal connection to them, or underestimating their abilities due to our cultural conditioning, perhaps the best approach would be an openness to being surprised by their abilities. Is it so wrong to assume an animal can think and feel, and look for proof otherwise, rather than to assume it cannot and require proof that it can?

So, here we are, early in the twenty-first century, with a world and culture built on the *scala natura*, but with a growing mass of evidence that the scale isn't so natural after all. Yet whether through religion or culture, we continue to view animals as creatures to be ruled over. In the British legal system, animals are considered "things"; some laws dictate how they are to be fed, sheltered, and killed, but they do not have legal rights like humans do, like the right to life. In London, where I live, it is normal to eat animals, use them for clothes, emotional support, and to cover furniture. Perhaps this is the legacy of our exceptionalism.

I once asked Roger Payne what he thought had been holding us back for so long from trying to speak to animals. "It's exactly like white supremacy, only it's human supremacy," he said, "and like white supremacy, it's based entirely on fear." I think he's right. We are right to be afraid of what we might discover. Giving up the privileges that you have enjoyed over others is a frightening thought. To communicate with another species would force a reckoning in how we have treated so many of them.

However, I suspect these fast-accumulating animal discoveries *are* starting to affect our cultures and our decisions, gradually and erratically. In the words of sci-fi writer William Gibson: "The future is already here—it's just not very evenly distributed." In fact, even a scientific consensus on one of the last great anthropodenial holdouts—animal consciousness—has been building. For instance, in 2012, a convention of scientists from a range of disciplines met at the University of Cambridge and issued "The Cambridge Declaration of Consciousness." It read: "The weight of evidence indicates that humans are not unique in possessing the neurological substrates that generate consciousness. Nonhuman animals, including all mammals and birds, and many other creatures, including octopuses, also possess these neurological substrates." Five years later, a report by the European Food Safety Authority—where seventeen experts reviewed 659 scientific papers—found "examples of higher levels of consciousness in domestic livestock." The report referenced studies showing that hens could judge their own state of knowledge, suggesting they were conscious of what they knew or did not know. That pigs could remember what events they had experienced, where and when. That sheep and cattle could recognize individuals. That among the most elaborate capacities there was evidence animals have knowledge of their own state, the capacity to know and deal with their own knowledge, and to evaluate the psychological state of their fellow animals, and that this could lead to some form of empathy. "Collectively," the report said, "these studies... clearly support the hypothesis that domestic livestock species are capable of complex conscious processing."

Among the general public, similar ideas of animals and what they could perceive have been taking hold so much so that in 2017, the most viral political news story in the UK was that the Conservative Party had voted down a law stating that animals were "sentient beings" and could feel pain or emotion. It was shared half a million times and there was such a public outcry that Michael Gove, then secretary of state for environment, food, and rural affairs, was forced to respond with a video reassuring people that the party wanted "to make sure Brexit delivers not just for the British people, but for animals too." It's unthinkable to imagine this story receiving such traction just a decade or two earlier. Four years later, the Conservative government brought the Animal Sentience Bill to be voted on in the House of Lords; this legislation recognized that vertebrate animals could feel pain and should be protected. A group of Conservative MPs lobbied for them to go further and include some invertebrates such as octopus and lobsters. The TV show *Good Morning Britain* pithily tweeted: "Animals officially have feelings. Is it time to stop eating them?"

In 2017, a case was brought before the New York Supreme Court on behalf of two captive chimpanzees. The primates' lawyers sought a writ of habeas corpus from the court to free them from their unhappy confinement. While the court refused, one of those who made the call, Judge Eugene Fahey, later wrote in his legal opinion that he had "struggled with whether this was the right decision." He believed this was not the end of the "profound and far-reaching" matter. The question would have to be addressed of whether nonhuman animals should be treated legally as persons or as they currently are, as property; as things, with no rights to liberty or otherwise. This "speaks to our relationship with all the life around us," wrote the judge. He continued, "In elevating our species, we should not lower the status of other highly intelligent species."

The lawyer who brought the case to court, Steven Wise of the Nonhuman Rights Project, is next looking to represent captive orcas in a West Coast dolphinarium. If only there were some way for his clients to express themselves, their suffering, and their desire for freedom, he told me. In legal terms, it would be revolutionary.

The fact that courts will consider these matters, the rise of vegetarianism and veganism, the keeping of pets as companions rather than tools, and the wider environmental movement are all indications of increasing empathy toward other species—the slow but steady erosion of our anthropocentrism. The more we learn about other animals and discover evidence of their manifold capacities, the more we care, and this alters how we treat them. As the songs of humpback whales evolve each year, our culture, too, is changing.

At the end of this investigation, I therefore think we have a choice: to continue to believe whatever we want about the inner worlds and communications of cetaceans and other species and project it onto them, or to make the effort of finding out what is really there. This matters because speaking is one of the last absolute supports of human exceptionalism, one of the few remaining things we believe only humans can do. It matters because our exceptionalism is dangerous to us, too. When we see ourselves as above or outside the rest of the living world and don't value other ecosystems and life-forms, we take them for granted and use them up. Ultimately it matters for our own self-preservation: To a large extent, our survival on this planet depends on recalibrating our conception of how human beings fit in among the other lives on Earth. As the Swedish environmentalist Greta Thunberg put it in a film we made together, "Because we are part of nature, when we protect nature, we are nature protecting itself."

In 2021, Ari's colleagues conducted an analysis of 321 tag deployments on baleen whales in Antarctica and discovered that they consume far more krill (small, shrimplike animals) than we previously thought. From these findings, they learned that in prewhaling times, baleen whales in the Southern Ocean alone ate 430 million tons of Antarctic krill a year. This is twice as much as *all the seafood* we humans catch annually today. When whales consume and then excrete their prey, they scatter their matter across the oceans, like gardeners spreading fertilizer on vegetable beds, driving the cycling of nutrients across whole seas. The implication of this is enormous: Whales are the linchpins of marine ecosystems. Today, however, as their numbers have not yet recovered from whaling, the study's authors estimate that whales currently cycle only a tenth of the iron that they used to—a cycling vital to sequestering carbon from the atmosphere and sinking it deep into the seas. Whales grow faster than trees. And when one dies and falls into the deep, it takes some thirty-three tons of carbon with it. We thought we were just killing whales, but we were killing the seas and the skies as well. The International Monetary Fund estimates the lifetime value of the average baleen whale, based on services useful to humans, as at least $2 million, giving the current global whale population a value of *$1 trillion*. But since the rise of human civilization, 80 percent of wild marine mammals have been lost. Each year we kill billions more animals, and silence their minds. And as we accelerate further into this mass extinction, with each species we lose, we lose forever the unique way in which it sensed and processed the world. Our human exceptionalism has cost us dearly.

When I think back to the discovery of Neanderthals, I feel their loss. I wonder if we will feel this way about whales in the near future, if they go extinct, only for us to later realize what has passed us by: a chance to speak with a fellow traveler in life on Earth. A chance to connect our minds and to see how we ourselves are perceived, by senses and brains far different from our own. In the words of Robert Burns, "O wad some Pow'r the giftie gie us / To see oursels as ithers see us!"

But I'd had enough of the philosophy of science and speculation. I wanted to feel, once more, what it was like to spend time in the company of these creatures, to communicate with them in the most primal sense. I wanted to swim with a humpback whale.

12

Dances with Whales

We shall not cease from exploration
And the end of all our exploring
Will be to arrive where we started
And know the place for the first time...
Not known, because not looked for
But heard, half-heard, in the stillness
Between two waves of the sea.
—*T. S. Eliot, "Little Gidding"*

It was early in the morning and the sea was slightly choppy, the hot sun dazzling on the water. I took my seat in the skiff and checked my gear. Fins, mask, weights, belt, snorkel. There were nine of us on the little motorboat, including two guides. We pulled away from the much larger mothership, the MV *Sea Hunter*, on which we had just spent the night, and headed out into the open sea. We were sixty miles from the nearest land, moored against a colossal coral reef that sits atop a submerged plateau called the Silver Bank in the Dominican Republic. It is a breeding ground for North Atlantic humpback whales. Our leader, Captain Gene Flipse, stood by the boat driver. We drove past the bones of a 1970s shipwreck; one rumor, Gene told us, was the boat was a drug smuggler run aground, fleeing justice. We scanned the horizon for the breaths of humpbacks. We were looking for stationary whales, calm whales, whales that didn't mind people. Gene explained that it was possible to be near whales only like this, since

Humpback whales and divers.

with a twitch of a tail, a whale will outswim an Olympic champion. The encounters that worked were those that happened slowly, where the whales knew where we were and did not change their behavior.

We'd been on the water for a couple of hours, squinting in the glare from sun and sea, when we saw them breathing among the extruding tops of the coral heads. These giant beasts would be almost hidden, were it not for their telltale ventilation, the haze of condensing air flushed from their hot lungs, visible for miles to creatures of air like us. They were about five hundred yards away to our starboard side, and the boat gently maneuvered its way through the reef to where they'd been. Everyone on deck waited in anticipation for the next rise of breath. There were no other species of whales here, so these had to be humpbacks. I felt my heart beat in my throat at my jugular, where it pressed against my wetsuit. I had seen a lot of whales since Prime Suspect landed on us, but always from within a boat or kayak.

I turned to my cabinmate on this expedition, a radiologist from Minnesota named Sean. His eyes were fixed and wide, focused on the hint of distant whales. He'd asked why I was here, and I told him what I'd discovered about the hopes for AI and pattern recognition in biology. He said he was using machine learning himself and described how AI now helped him spot tumors in mammograms. They'd had the technology for only a few years. He said the machines sometimes found subtle hints of cancer that he missed, life-or-death patterns in pictures. Sean liked the machines, he said. They worked.

Our boat inched closer to the whales. There were two of them, resting together to breathe on the surface and then sinking back down into the water column. As they disappeared underwater at the end of their last breath cycle, their passing left a pair of flat patches of calm water on the surface. These were their "footprints." Gene eased himself off the boat and swam over to a footprint. When he got there, his arm shot up, indicating that he could see the whales. A few minutes later, his hand beckoned the boat over, which indicated that the whales were at ease. Sean, the other five whale nuts, and I readied ourselves on the side of the boat. I'd washed out my mask and sat above the water, watching Gene's hand for the final signal. Eventually, he beckoned to us again and I lowered myself off the boat, careful not to splash. After a brief, beautiful moment of vertigo, looking down into the endless blue, I kicked over to him, my breath fast, rolling my body so the kicks didn't break the surface as I swam. I reached Gene and looked down past his long fins, and there they were. Despite the murky visibility, their bright white pectoral fins shone through the deep like moths' wings, much lighter in pigment than the black and dark blue fins of their Pacific cousins.

It felt unreal, and yet so straightforward. There were whales, and here I was, bobbing on the surface gazing down at them. I felt nagging jabs of fear, fleeting thoughts about how stupid this was. I could already see the headlines: "Man Who Survived Leaping Whale Killed by Another Whale." Why go looking for trouble? Yet these anxieties were quickly overwhelmed by a feeling of awe. I was transfixed. These whales had come here from their feeding grounds in the North Atlantic, many from the Gulf of Maine and Bay of Fundy, others from farther: Newfoundland, Nova Scotia, Iceland, and Norway. Now that the ice of the pole is melting, the whales are spreading up into northern Russia. I thought of their long journey here, and my long journey through what we know of their minds and lives to join them.

The pair comprised a resting female and her male escort. Females often have "escorts" at the breeding ground, males that stick with them and try and keep other males away. It's surprisingly hard to tell a male from a female humpback at first; to be sure, you need to check the position of its privates. Gene pointed to one of the whales, resting lower than the other, which had what looked like divots down its back, white with scar tissue. Gene's theory was that they were birth scars. That in birthing their one-ton babies, sometimes females end up pressed against submerged landforms, which here are razor-sharp coral heads. The scars marked a calving, a mystery no human has witnessed in its entirety, despite humpbacks being among the most studied of all the cetaceans. I floated gazing at this whale and thought of this scene, of how she might have carried her calf a year inside her before giving birth underwater in the dark, pressed quaking against the living rock.

Gene looked at his watch. They'd been down ten minutes, he said. Somehow, I'd already been hanging next to him for five. I hovered above the whales on the surface, as if tethered like a kite above two zeppelins. I listened to my breathing and thought of theirs, of how weird it was that these had been land animals. Then I thought how weird it was that all our ancestors had once been sea animals. I started to hold my breath to see how long I could match them. I looked at the humpbacks hanging unmoving beneath me in the dark blue, parked in the current like two buses in a night depot. Perhaps we weren't as different as we looked. I ran out of breath and exhaled. They continued, hanging there.

Lying facedown, the sun on the back of my head, I ate their bodies up with my eyes, taking in all the details. The scars, the scratches, bite marks on their tails from orcas, mottled light and dark pigment. Their eyes in their bulges were just visible. Surely, they could see us, I thought, they knew we were here. I drifted in and out of a sense of scale vertigo, my brain turning over and over how big they were and trying to compute it. For twenty-eight minutes we hung above the whales and breathed while they did not.

The bull had seemed settled, but started to shift position slightly above the female. He hovered above her for a few minutes and then made a sudden movement with his pec fins and sped off ahead into the gloom. (This sounds strange for such a big beast, but he did!) When we'd seen this before, it generally heralded the arrival of another male—but this time, we could see no whale approaching the pair. What happened next was more worrying. The female he'd left behind turned her nose toward the surface and arched her back like an athlete preparing to do a forward flip. She slammed her tail flukes down, propelling herself upward toward the surface. In one movement she had covered ten yards, before she swept her tail up and down once again. This took less than two seconds.

I felt a strangely familiar disconnection from myself. *Not again*, I thought. She was going to breach. The female propelled upward, now two body lengths away, and with relief I realized she was moving away from us. The change from total stillness to thrust was almost unreal in its power, an airship transformed into a biomechanical drag racer. She pulled her pectoral fins back as her body performed a handbrake turn that brought her angle of attack to the vertical. I briefly remembered a visit I once made to a Florida swamp to watch a space shuttle launch. The explosion of the ignition changed a lump of metal sitting on a tarmac to an inevitable missile, and so it was with this whale. One last great twitch from her peduncle sent the enormous cartilage propeller of her tail fluke carving through the water, and up she went.

I watched her break the surface as she arced into the air, the sun glinting off her skin, water pouring from her sides—peeling away like the water vapor condensing off the Saturn V rocket as it takes off in old NASA footage. She flipped and rotated as she flew, a thirty-ton reverse backflip twist. I stuck my head back under as she landed to see the whitewater of her impact and heard its *whump* in my ears: the apotheosis of the most beautiful animal movement in the world. It was an embodiment of power. It must have looked like this from beneath our kayak in Monterey, the flick of a mental switch, the engagement of one of the biggest muscles life has ever produced, and voilà—a whale can fly. For a moment.

Water gushed into my mouth around my snorkel and I realized that I had been grinning wildly. Had I come for this? It brought back memories of sudden dark plunging and near death, but it was simpler than that. I felt awe and I felt joy. I had come halfway around the world to a remote patch of sea to feel tiny and unknowing, and it was fantastic.

Everyone whooped and yelled and the whale continued breaching, twice more, farther and farther away from us. The boat and its anxious driver swiftly came to sweep us up. Jeff, the guide, had spent half his life in the water with whales. For thirty years, he'd spent four or five months of the year swimming with humpbacks twice a day. He told me he'd only ever seen five breaches from the water. He looked at me sideways. "Something is weird with you," he said.

Ari Friedlaender estimated for a humpback to fully breach it would require 9.8 megajoules of energy, with the whale at its maximum generating fifty kilowatts of power, which is enough to power a home for a day. Whales will breach again and again and again. That we still do not know why the biggest animal does this—communication, display, parasite removal, none or all of the above—is as good an example of our ignorance about the living world as any other. Had she shaken off her suitor and called another over? Was she making her feelings known about the annoying bobbing humans? Had the leap come in response to some call beyond my hearing?

I felt, very simply, lucky. Not just to have survived the whale, and to have been on the right side of this one, but to be alive at a time where there are leaping giant whales to survive. As the other tourists pulled off their fins and belts and hauled themselves aboard the boat, I waited and lay back in the water. I shivered with adrenaline, and with a feeling like I was watching Michael Jordan deliver a slam dunk, the pure thrill of seeing someone do something so well, something they were born to do. I guess I should have felt surprised to see another breach so close, but I didn't. I felt a strengthening of the truth that the more you look, the more surprising things you'll see. That if you look at something that is not much observed, then all is unexpected, because our imaginations fall short of the lives of these animals. The only surprise is that it took a whale jumping on me to start me looking. This was not closure to my near death but rather an extension, a return, a way of seeing it again but with new eyes.

The Silver Bank is a rocky underwater shelf sixty-five nautical miles off the north coast of the Dominican Republic. It rises from the deep-sea floor to a hundred feet beneath the surface of the water. Poking up out of the water at

A humpback comes to have a look at the author.

points are coral reefs. Bane of sailors, studded with wrecks, it is named after a Spanish galleon laden with silver that sank here 350 years ago, lost on the human migration route of wealth from the New World to the Old. In the 1970s, the forward-thinking government of the Dominican Republic created a giant marine sanctuary off the north of their island (Hispaniola). To protect the whales but support the whale-watching industry, only three boats are issued with permits to take people to the Silver Bank to swim with the whales. Once there, they must abide by strict rules, and will remain within an area comprising less than 1 percent of the reserve. This ecotourism funds the protection of the whales with minimal disturbance.

These whales had journeyed to this reef and back from the cold northern waters, some even coming from the Arctic, every year for the entirety of their lives. Their parents and ancestors before them had done the same, for perhaps hundreds of thousands of years. Over the past few decades, they have been shadowed in both breeding and feeding grounds by strange new creatures: humans. Massing in their tens of thousands, often outnumbering the whales themselves, whale watchers like me come, many traveling even farther than the whales to reach Antarctica, Hawai'i, Stellwagen Bank, Monterey Bay, Alaska, Australia, Russia, Mexico, Norway, Sri Lanka, and South Africa. A breeding ground, the Silver Bank is home to the greatest concentration of humpback whales in the Atlantic. They come here to sing and calve and breed.

I was intrigued by those who had come to witness them: my fellow passengers. Sean, the boyish fifty-year-old radiologist from the Badger State, was my bunkmate. He'd come before, but his previous trip was thwarted by such unseasonably foul weather to the degree that they had only made it out of the harbor after four days, with his wife so seasick by this point she vowed never to come again, and Sean so hyped he immediately booked to return. Also aboard were a rambunctious English extended family, who had decided all their holidays would be adventures; a loved-up couple in their fifties from the San Francisco Bay Area; a woman from New Jersey with her aunt; a lone, introverted German woman adrift among the English speakers; and Jodi Frediani, a friend in her seventies I'd met filming on whale-watching boats in Monterey Bay. It seemed an absurd coincidence that Jodi was aboard until I asked her how many trips she'd been on. This was her fortieth. "I don't drink or do coffee," she said, her eyes twinkling. "This is my addiction." It was an expensive addiction—I'd burned through a chunk of my savings—but boy, was it worth it.

What drove these people? Gene told me about how he'd once been at sea and encountered six adult sperm whales. As he watched them, the water around one, a female, had darkened with a great flush of blood. He had no idea what was going on, but then saw a newborn calf appear out of the "bloom of red." Sperm whales have the longest gestation of any animal, longer than even the humpback whale: eighteen months, culminating in a brutal birth. He said that while the mother lay recovering, one of the males looked after the infant. Whales are conscious breathers: They must think to get their blowholes above water and inspire, and this is a lesson newborn whales must learn quickly, or drown. Gene told me about how the colossal bull gently tended to the baby, nudging the calf to the surface to take its first breaths. Perhaps the bull was its uncle or brother. He'd watched for twenty minutes and assumed the whales had not minded the little boat he was on, so he slipped into the water. He'd swum out alone and the male had turned and checked him out, his sonar scanning the visitor. It rang Gene's lungs. He said it felt like someone flicking you in the chest with their finger.

Gene has since spent his life swimming with giant ocean beasts. He has danced with whales, where man and whale would take turns mimicking one another's movements. Sometimes this dancing mirror act would last an hour. But he told me if I wiped his mind to leave only one encounter, it would be his swim with the sperm whale bull as it tended the calf.

While on the expedition, I was reading Robert Macfarlane's book *Mountains of the Mind*. In it, he writes of the eighteenth-century thrill-seekers who traveled across Europe to be confronted by what they called "the sublime." Mountains, volcanoes, and glaciers, landforms of colossal size, cruel weather with the power to smudge out human lives. Geological features that were themselves recently discovered to be fragile and ephemeral when viewed across the other discovery of deep time, newborn mountains ground into dust and silt over just a few hundred million years. Enormous but still themselves fleeting, these sublime landscapes cast human existence into small and pitiful relief. These were places where people could feel their mortality against raw geology, as you can test a knife's sharp edge against your soft thumb. The early explorers of the mountain landscapes found they had no words to convey what they were like, what they looked like, what they felt like, because there was nothing they could compare them to that would make sense for the readers of their letters, who had never

The eye of a humpback whale.

left the plains. Still today, Macfarlane writes, they "challenge our complacent conviction—so easy to lapse into—that the world has been made for humans by humans... They induce, I suppose, a modesty in us."

His words struck a chord. Those who venture here, like those long-dead junkies of the sublime, come to feel indescribably small. But they don't just come to look at whales; they come to be beheld themselves. Unlike mountains, whales can look back. To be gazed upon by an animal so big, which you hold in such esteem, is for many a transcendent experience, one worth traveling thousands of miles and spending your life savings on. After swimming with a calf one day, where the infant whale had returned again and again, seemingly playing with her human visitors (all the while carefully observed by her resting mother below), I turned my attention to the other people. Sean was beside himself. "Did you see?" he said. "It looked right at me. With its eye, it looked right at me and it saw me." Jodi paddled past me with her camera, clasping the giant glass dome to her chest like an infant with a faraway look. "See, Tom, this is why I keep coming back." We had been noticed by something incomprehensible and enormous. It was sublime. As Macfarlane writes about mountains, "True, you learn yourself to be a blip in the larger projects of the universe. But you are also rewarded with the realization that you do exist—as unlikely as it may seem, you do exist."

There was one encounter I had come in hope of more than any other. And one morning during my trip, it happened. We found a singer. We slipped off the boat and I swam over to Gene and found I was in water ringing with song. The wind was up and the visibility, normally more than thirty yards, was unusually poor. Below Gene in the silty blue, I could just make out the white pec fins. The whale was hanging vertically, flukes up, head down, like an ogre in a snowstorm. It sang. Humpback whale song comes out louder on the fronts of their bodies. We maneuvered ourselves around into the full stream of song. It was like pressing yourself to the speakers at a rave. My lungs and airspaces and limbs all vibrated, and I felt I had become the medium it was talking in. As my legs tingled, I thought of their jaws, which, as Joy Reidenberg told me, they use to catch the sound waves of songs and channel them to their ears, and I wished I could listen to this song as a whale does. It was an experience at once almost religious and yet absurd, the grunts and tiny squeaks this enormous beast produced like jazz composed of seals and bagpipes, creaky doors and ghosts. Some were almost like the yells of happy people from the far end of a cave, some like the rumble of indigestion, others like wails. I found myself making little noises of delight into my mask. The whale came up for breath and the song paused. As he descended and took his place, he picked up his tune and went through the same song for half an hour. Then he rose again and returned to repeat it once more. The longer I hung there, the more I began to recognize patterns. The most obvious was at the end: The song had a distinctive finale, the same each time. He had swum so far to sing this song, which would change next year, an unrepeatable performance. What did it mean?

I thought of how far we'd come, in the lifetime before my own, moving from mass slaughtering to the understanding that whales sing, and putting their songs on spacecraft—and I thought of what more we might discover before I die. As a convert to the possibility that cetaceans might talk, I wondered what patterns we might find in their voices. Do humpbacks, like dolphins, have words for themselves and their groups? What will quantum AI, when it comes of age in a decade, find in Julie Oswald's dolphin data? What will CETI's sperm whale expeditions reveal, and what constellations of click phrases will they chart? What evidence for their capacities will Joy Reidenberg and Patrick Hof and their scanners unearth in their soft brains? With ever more interested humans, with ever more sophisticated devices recording cetaceans, what revelations will be pulled by Ted Cheeseman and programmers like Jinmo Park from the shared big data? If we have learned so much in such a short time, and our capacity is growing so fast, could the moment come when we can hazard a guess at how to say hello in North Atlantic humpback whale dialect?

I hung in the sea ringing with the voice of a beast composed of flesh and bone and mystery, and I wondered who would come next to change how we see these animals, and how we interact with them. And yet in that moment, all I could feel in the water was that it didn't matter if whales weren't telepathic or if they didn't have all of Hockett's elements of natural language, if their brains did not confer on them consciousness like ours. What matters is they are there.

* * *

Life on Earth runs on code. You and I and everyone who has ever lived was constructed from and could be reduced down to genetic instructions written in efficiently contorted protein chains called deoxyribonucleic acid, or DNA.

There are many ways to represent a person. You can take a photograph of your lover, capturing a signature of how light bounces off them in a digital reflection of zeros and ones. No human can read these, but printers can, and they can use this code to tell the ink heads they pass back and forth over a paper when to lay pigments upon it, making a two-dimensional picture of your lover, recognizable to anyone who has met them: a portrait. You can look at this picture years later and see your lover's mood, their age, their health, their gait. But they are encoded in other ways, too. You can take one of their hairs or skin cells and extract their DNA. You can use machines to transcribe this into a series of letters ATCG, representing the fundamental units of the code that makes a person, the base pairs. You can print out the code for a human represented as base pairs and bind it as a book, though no human could read it. But you can put it into a molecular machine, like a human egg, and the DNA code will tell the egg how to divide and change itself to form your lover cloned anew. We are all made of code; our bodies rely on it to speak to themselves, as does our technology.

We also rely on codes for signaling to others: Bacteria, trees, humans, coral reefs, marmosets, and earthworms all send out codes via electrical flashes, pheromonal clouds, sounds, movements, and invisible chemical trails. We all send out vital information in forms that some can read and others cannot.

In 1990, a hugely ambitious international scientific expedition was launched. It wasn't aimed at the stars, or at the deep sea, but internally, at the terra incognita of our own DNA. It was called the Human Genome Project (HGP) and its goal was to map all the genes in our species, *Homo sapiens*. We had only recently learned the structure of DNA in the 1950s. After this, scientists had found out that there seemed to be patterns in it, made up of simple repeating units. The arrangement and sequence of these units was a code that made a person. At first, we explored this genetic code by mounting inward expeditions bit by bit, looking for particularly exciting or accessible sections of DNA to foray into and map and describe. We reached for low-hanging fruit.

The HGP wanted to map the *entire* code, all 3,000 million bases of it: the whole genome. With this total overview we would have, for all of humanity to explore, the code for making a human being. A multidisciplinary effort that brought together teams from chemistry, biology, physics, ethics, engineering, and informatics, it remains the largest-ever collaborative biological project. It ran for thirteen years at a cost of $5 billion—and it worked. It revolutionized how everyone approached trying to understand and manipulate the human genetic code, encouraged and connected a generation of scientists in a shared goal, and transformed the costs and practicalities of carrying out genetics work. Today it costs just $1,000 to map an entire human genome.

Two hundred years ago, your great-great-great-great-great-great-grandmother could have looked at her child and wondered what was responsible for the pattern of hair and skin and bone that made them, and she would have been in the dark. Today, thanks to the field of genetics, she'd be able to explore her ancestry, as well as that of her child, and could even learn how that child might look half a year before they emerged from her womb. But if you were to watch a flock of crows gather, the patterns of calls and movements of response, you could find out little more about what they mean than your ancestor could. Indeed, she may have paid them more attention, and been able to read their codes far better than you can today.

Imagine what an HGP for animal communication might reveal. Our communications, like our genes, are the results of evolution: The same pressures shaped them, and likely we have much in common with our animal relatives. By sequencing the chimp genome in 2005, we could compare its patterns of genes with our own. We discovered that we share 99 percent of our DNA with our close cousins. We share 85 percent with a mouse, 90 percent with a cat, 61 percent with a fruit fly, and 40 percent with a banana. By seeing how we fit in, we better understood our place, and our true differences and similarities, not those projected by how different we would like to be. There was an outcry at the discovery of how similar our genetic instructions are to those of chimpanzees, and many people disputed and dismissed the findings. It is perhaps uncomfortable to brush genetic shoulders with a beast if you consider yourself incomparably different. Now with the HGP we can add other genomes, compare and see where we differ, and which mechanisms we share. We can find out what these differences mean and what they do.

Imagine if as well as a renewed space race between rival billionaires, the same firepower and finance, the same international spirit of endeavor and competition, were to be trained on decoding the messages of what are, as far as we know, the only other sentient lives in the universe—those on Earth. Imagine spending $5 billion to map the communication of another species and compare it with human communication. Imagine making first contact with another sentient species on Earth. If you are so inclined, imagine the everlasting fame. For now, animal communication researchers are like pre-HGP geneticists, working in many small unconnected teams forced to focus on decoding the simplest bits of animal communication, or those whose funding is easily justifiable. We have no overview and no map. Those working are aware only that what they are charting are the nearest, brightest, most easily accessible stars in a universe of encoded animal signals, with the majority of people unaware there are any stars there at all.

The Human Genome Project, the Manhattan Project, the Apollo program. These endeavors amassed the finest minds and enormous budgets and aimed at moonshot moments that unlocked unparalleled learning about ourselves, the forces of nature, and our place in the universe. The Apollo astronauts sped to the moon, where they found dry, old gray dust, and then flew back. Imagine if they had known that on the moon were the surviving remnants of a great civilization that soon may be gone forever, perhaps our only company in the story of life, and that there existed a chance, no matter how slight, to communicate with them. But we don't need to go to the moon to attempt this. We have forgotten the wonder of the other animals we live with because they are among us, and because we have gotten more used to thinking of them as resources we can simply use. Of course, not every van Leeuwenhoek peers into what they thought was just pond water and finds it alive with animalcules. Not every Hubble trained on empty space finds it full of galaxies. But you cannot know until you decide to look.

Author and humpback.

Over the years, since coming round to believing in the possibility of animal communication, I have, of course, bugged Britt and Aza with the unanswerable question—one you may also be asking at the end of this book: *So, when will we be able to speak to animals?* They, along with everyone else working in this field, have no response for now. But using a longer timescale, I asked them what they felt could reasonably exist in the world by the time my unborn daughter was our age, in 2055, and here's what they said:

> Nature documentaries could be subtitled. Ships can speak to whales, dolphins, orcas, and other marine mammals to let them know of our approach, reducing deadly ship strikes to a minimum. New perspectives of what it means to be alive, to love, to live on this shared planet, are integrated into human culture, changing the perspective of ourselves and our identity as a species. We learn that we are not alone in the universe. We gain deep new insight into the plural nature of consciousness.

Reading this put me in mind of the words of the great Mark Twain: "For the majority of us, the past is a regret, the future an experiment." Our past with these giant animals has indeed been regretful. I'd like us to make our future a hopeful and ambitious experiment.

Perhaps soon enough there will be a breakthrough that later we will point out as significant. An app that analyzes your dog's face will become huge and the vast revenues of the pet industry—which is now comparable in size to the arms industry—drives a revolution in animal-decoding tech. DeepMind or Open AI will decide that cracking a two-way conversation with a dolphin is its next aim, bringing its colossal human expertise and computing power. A tool kit of user-friendly, general-purpose AIs is made available to biologists and citizens and spreads around the world, harvesting patterns from the world around us on unprecedented scales. And can those humans who share this endeavor resist pressures to keep their discoveries secret, their data closed off, to hoard funding and to hog acclaim? Will the better angels of our nature guide us as we decode nature?

What is certain is we will continue to find patterns in nature, and we will continue to be surprised by learning that other species can do things we previously thought our own unique preserve. But while our tech develops, as our inclination to look deepens, as our understanding of how little we have found increases, as we see more and our questions proliferate, will this be in step with the destruction of what we are studying? To be alive and explore nature now is to read by the light of a library as it burns. Could our discoveries prompt us to put out the flames? The truth is that you and I, the ones alive right now, will see.

I do not watch the sea like I used to, before I set out on the journey of this book. Before, I would just take in the view. Now, however, my eyes skip around it, scrutinizing the shape of spray, a white broken wave where there's no rock or wind. A flash on the horizon suddenly reveals itself to be the glint of a fin catching the sun; every stirring of the surface interrogated in the hope that it is the clue to a whale underneath. One afternoon I was looking at the sea, my wife, Annie, six months pregnant beside me with our daughter, still for now an aquatic being in her womb, yet to meet the air. I scanned the waves, reassuring myself there were no hidden cetaceans.

And then I thought: What if there were none at all? What if every splash everywhere was just a splash, if no more fins broke the surface? My stomach turned over on itself. They face a troubled future. Some species are going extinct, right now. I want my daughter to live in a world where these creatures thrive, across the seas in all their forms. Where their cultures evolve and shift and mingle, and their strange voices fill the depths. I want this world for them, but also for her, for what she might gain from their wild influence, from the things we are on the cusp of knowing about them.

My daughter will surely grow, and I will surely age, and whether it is from another leaping whale falling on me or I trip on the stairs, I will die, and she will have to learn what it means to lose something forever. This is inevitable. But there are losses we do not have to learn to accept, which we can choose to stop. The fates of the whales and dolphins are in human hands, and this loss is one I do not want for her. I hope that when she looks at the sea, an aged woman, she will catch a glimpse of a leaping spinner dolphin or humpback whale, and perhaps when she sticks her head beneath the waves and hears their whistles and songs as I did, they will mean something to her. And perhaps, just perhaps, she will be able to answer back. *I am here,* she would say, *you are here and I am here.*

ACKNOWLEDGMENTS

A dolphin cocktail party in the surf.

Like many whales, I am a social beast. And like the song of the humpback whale, this book only exists thanks to my fellow animals. I take this chance to recognize you who have helped me in this wonderful task.

Thank You, In (Sort of) Order of Appearance

Thank you, the cetacea, if you weren't so magnificent, I would have nothing to write about.

Thank you in particular, Humpback Whale CRC-12564. Thank you for not squishing us so spectacularly and gifting me with both the story of a lifetime and a handy author platform. I don't know your name in Humpback, sorry we called you Prime Suspect.

Thank you, Charlotte Kinloch, for your humor, patience, courage, and wondrous webbed toes. I am so glad we were nearly annihilated together.

Thank you to all the scientists and whale people in this book who gave me so much of their time and help, and trusted me to tell their stories. I hope I did it justice. Thank you also to all the scientists and whale people who did the same but who I couldn't squeeze into this book. I tried my best, and when I started, I thought I'd be able to, but then my first draft was 140,000 words long and I got in trouble. I salute you for going out of your way to help me: Tania Howard, Michelle You, Sabena Siddiqui, Hazen Komraus, Ru Mahoney, Dave and Pat Allbee, Nancy Rosenthal, Hartmut Neven, Holly Root-Gutteridge, Julie Oswald, John Ryan, Joy and Bruce Reidenberg, Steve Wise, Jodi Frediani, Peter Read, Colin Burrows, Wesley Webb, Mike Brooke, Marie Phillips, Gene Flipse, Roger Payne, and many more who my leaky memory betrays. My sails were filled with your goodwill.

Apologies to all the invisible scientists, the creators of all the discoveries presented in this book. If I wrote all your names in the text it would be unreadable, but this does feel unfair. I salute you; none of this knowledge would be here without your work.

Thank you, Kerry Glencorse, my agent, who before she was my agent thought this weird idea sounded like a good book, and encouraged me to see if I could write something. I could not have wished for a better guide. And Susanna Lea, a woman equally at ease swimming with sperm whales and the big beasts of U.S. publishing.

Thank you to my superb (and dishy) editor at William Collins, Shoaib Rokadiya, whose enthusiasm and vision for this book has somehow lasted undimmed from our first meeting, sustained me throughout, and developed into a friendship forged in chilly seas. Sho chewed into my bloated carcass of a manuscript with the relish of a pod of killer whales setting into an elephant seal. I'm sorry (not sorry) for trying to sneak in so many extra thoughts, chapters, factoids (still trying).

Thank you to my brilliant editor at Grand Central, Colin Dickerman, who scanned what I thought was my finished manuscript as a dolphin mother scans its newborn calf, with love and thoroughness. And like a dolphin mother he nudged this wobbly writer firmly to the surface to take his first breaths. Colin, I salute your rigor, and I have no idea how you reply to emails so quickly.

I thank you, eagle-eyed copy editors: Madeleine Feeny and Mark Long. Lord knows I gave you a lot of work to do. At HarperCollins (from whose London offices you can see hunting peregrines!) I was well looked after by Alex Gingell (project manager), Jessica Barnfield (audio), Helen Upton (publicity), and Matt Clacher (marketing). I hope you all caught some whale fever in the process. Jo Thomson made the mesmeric cover design for the UK edition. At Hachette, thank you to Rachael Kelly (editorial), Stacey Reid (production), Kristen Lemire (managing editorial), Tree Abraham (art), and Matthew Ballast (publicity). Thank you, Ksenia Dugaeva, for sleuthing the picture rights, and Andy Nixon, for fact-checking it. Turns out I'd spelled "anthropodenial" wrong the whole way through. That would have been embarrassing.

Thank you to the humans with nothing to do with this book, but who have enthused and supported me along the way. Sea brother Sam Mansfield, for the seal swims. OK David, for the fruits of your unusual brain. Steve Floyd, for being tickled by sea cucumbers. Sam Lee, Grant Jarvis, and the Lears, for the forests and song. Cherry Dorrett, for the wise counsel. Big Chris Raymond, for pizza, frisbee and joie de vivre. Brother Ollie for internet skillz. Mother-in-law Jenny Shaw, thank you for your consistent encouragement and for your astonishing painting of being underwater with the whale. Father-in-law, best-selling author Richard Wilkinson, thanks for your gentle inquiries as to how the book was going, even though they always seemed to coincide with when it was not going! Sabrina, sister from another mister, queen of emojis, who encouraged me to try to write this even though it meant I might abandon making films with her; Hampus, who I hardly know but still kept emailing me to tell me to write the book and even sent me a book about how to write a book. Thank you, Ian Hogarth, my friend, who scared the crap out of me when he told everyone I cared about in the world, in his best man's speech, that the book was going to be great before I'd even started writing it. Harry Birtwistle, godfather. It was done in time for you to read it. Your last gift to me, your kind words. I miss you, Harry.

I didn't know how to write a book before, but I was fortunate to have been taught storytelling by my apprenticeship in my television family. David Dugan, Andrew-Graham Brown, and Hugh Lewis, thank you for teaching me stories and emotions when I was obsessed with pictures and facts.

Thank you to the atmosphere. I am sorry that in making this book I added more carbon dioxide to you when that's the last thing you need. I worked hard to minimize this. I've paid for Supercritical to sequester the book's footprint.

The Law of the Tongue must be honored. Ten percent of the profits from this book will go to whale conservation. If each whale takes out thirty-three tons of carbon in its lifetime, and this book helps protect them, hopefully the whales and climate will come out on top.

In the book I use "we" often. I don't mean all humans; I mean people in my birth culture. Apologies to all those who come from those cultures and societies where traditional ecological knowledge already teaches what we in my homeplace are so slow to hear.

Thank you, Dmitri Grazhdankin, who changed the course of my life as much as the whale did. Dima, you introduced me to the pleasures of Stanisław Lem, helicopters and black tea, and taught me more around Siberian campfires about the process of scientific discovery than I learned in three years at Cambridge.

Thank you, Dad, for teaching me the importance of making complicated things clear, and the encouragement to just *go for it*. You will never read this, but your mind runs through it all. Thank you, Mum! When I was ten, I wrote a short story and you liked it, and I remembered that whenever I found this hard. I love you.

Thank you to Stella for giving the book its ending, and my life a new beginning. You are wonderful.

Thank you, my wise and fabulous wife, Annie, for providing the world with Stella, and for helping me all the time with everything, including loads of this book, and for always getting in the sea with me, and laughing, even when there are sharks.

No Thanks To:

SARS-CoV-2. You awful sack of RNA. Just go away.

NOTES

Epigraph

vii **"How do you expect":** Stanisław Lem, *Solaris* (London: Faber and Faber, 2003), 23.

Introduction

xiii **"What if I had never seen this before?":** Rachel Carson, *The Sense of Wonder: A Celebration of Nature for Parents and Children* (New York: Harper Perennial, 1998), 59.

xiv **it was hard to make anything out:** Paul Falkowski, "Leeuwenhoek's Lucky Break," *Discover Magazine*, April 30, 2015, https://www.discovermagazine.com/planet-earth/leeuwenhoeks-lucky-break.

xiv **275 times:** Felicity Henderson, "Small Wonders: The Invention of Microscopy," *Catalyst*, February 2010, https://www.stem.org.uk/system/files/elibrary-resources/legacy_files_migrated/8500-catalyst_20_3_447.pdf.

xiv **more than five hundred microscopes during his lifetime:** Nick Lane, "The Unseen World: Reflections on Leeuwenhoek (1677) 'Concerning Little Animals,'" *Philosophical Transactions of the Royal Society B: Biological Sciences* 370, no. 1666 (2015): 20140344.

xiv **comparable to those of modern light microscopes:** Michael W. Davidson, "Pioneers in Optics: Antonie van Leeuwenhoek and James Clerk Maxwell," *Microscopy Today* 20, no. 6 (2012): 50–52.

xiv **"animalcules":** Antony van Leewenhoeck, "Observations, Communicated to the Publisher by Mr. Antony van Leewenhoeck, in a Dutch Letter of the 9th of Octob. 1676. Here English'd: Concerning Little Animals by Him Observed in Rain-Well-Sea. and Snow Water; as Also in Water Wherein Pepper Had Lain Infused," *Philosophical Transactions (1665–1678)* 12 (1677): 821–831.

xv **"all huddling and moving":** Antonie van Leeuwenhoek, letter to H. Oldenburg, October 9, 1676, in *The Collected Letters of Antoni van Leeuwenhoek*, ed. C. G. Heringa, vol. 2 (Swets and Zeitlinger, 1941), 115. www.lensonleeuwenhoek.net/content/alle-de-brieven-collected-letters-volume-2.

xvi **"most ingenious book that ever I read in my life":** Samuel Pepys, *The Diary of Samuel Pepys*, edited with additions by Henry B. Wheatley (London: Cambridge Deighton Bell, 1893), entry for Saturday, January 21, 1664. https://www.gutenberg.org/ebooks/4200.

xvi **"exceedingly curious and industrius":** "The Unseen World: Reflections on Leeuwenhoek (1677) 'Concerning Little Animals.'"

xvi **"fairy tales about the little animals":** Letter from Leeuwenhoek to Hooke, November 12, 1680, in Clifford Dobell, trans. and ed., *Antony van Leeuwenhoek and His "Little Animals,"* (New York: Russell and Russell, 1958), 200.

xvi **"surprized at this so wonderful a spectacle":** "Hooke's Three Tries," Lens on Leeuwenhoek, www.lensonleeuwenhoek.net/content/hookes-three-tries.

xvii **better than a human can:** David L. Chandler, "Is That Smile Real or Fake?" *MIT News*, May 25, 2012, news.mit.edu/2012/smile-detector-0525.

Chapter 1. Enter, Pursued by Whale

1 **"the hottest blood of all":** Quoted by Captain James T. Kirk in *Star Trek IV: The Voyage Home* (Hollywood: Paramount Pictures, 1986). Quoted from D. H. Lawrence, "Whales Weep Not!"

1 **deeper than the Grand Canyon:** "Monterey Canyon: A Grand Canyon Beneath the Waves," Monterey Bay Aquarium Research Institute, https://www.mbari.org/science/seafloor-processes/geological-changes/mapping-sections/.

2 **Blue Serengeti:** Tierney Thys, "Why Monterey Bay Is the Serengeti of Marine Life," *National Geographic*, August 12, 2021, https://www.nationalgeographic.com/travel/article/explorers-guide-8.

5 **"stable multiyear associations":** Christian Ramp, Wilhelm Hagen, Per Palsbøll et al., "Age-Related Multi-Year Associations in Female Humpback Whales (*Megaptera novaeangliae*)," *Behavioral Ecology and Sociobiology* 64, no. 10 (2010): 1563–1576.

7 **Scientists have estimated the forces involved:** Paolo S. Segre, Jean Potvin, David E. Cade et al., "Energetic and Physical Limitations on the Breaching Performance of Large Whales," *Elife* 9 (2020): e51760.

8 ***allowing it to sink*:** Jeremy A. Goldbogen, John Calambokidis, Robert E. Shadwick et al., "Kinematics of Foraging Dives and Lunge-Feeding in Fin Whales," *Journal of Experimental Biology* 209, no. 7 (2006): 1231–1244.

10 **one hundred thousand views:** Sanctuary Cruises, "Humpback Whale Breaches on Top of Kayakers," YouTube, video, September 13, 2015, https://www.youtube.com/watch?v=8u-MW7vF0-Y.

11 **"I think you two survived":** Joy Reidenberg, email, September 18, 2015.

13 ***Time* magazine:** Megan McCluskey, "This Humpback Whale Almost Crushed Kayakers," *Time*, September 15, 2015, https://time.com/4035011/whale-crushes-kayakers/.

13 **"did you realize it was a whale?":** *BBC Breakfast*, BBC One, TV broadcast, February 9, 2019.

14 **lit by underwater torches:** Manta Ray Advocates Hawaii, "Dolphin Rescue in Kona, Hawaii," YouTube, video, January 14, 2013, https://www.youtube.com/watch?v=CCXx2bNk6UA&t=9s.

15 **pushes her along with its fin:** BBC News, "Whale 'Saves' Biologist from Shark—BBC News," YouTube, video, January 13, 2018, https://www.youtube.com/watch?v=2xMLwAP2qyk.

15 beluga chirruping and squeaking along: Simon Houston, "Whale of a Time," *Scottish Sun*, November 8, 2018, https://www.thescottishsun.co.uk/news/3464159/journalist-beluga-whales-50million-viral-sing/.

15 A beluga whale came to live in the Thames: Matthew Weaver, "Beluga Whale Sighted in Thames Estuary off Gravesend," *Guardian*, September 25, 2018.

15 I secured a commission to make a film: *Natural World*, season 37, episode 7, "Humpback Whales: A Detective Story," Gripping Films, TV broadcast, first aired February 8, 2019, on BBC Two.

17 new species of mammal-hunting orca: Douglas Main, "Mysterious New Orca Species Likely Identified?" *National Geographic*, March 7, 2019, https://www.nationalgeographic.com/animals/article/new-killer-whale-species-discovered.

17 a mysterious new deep-sea whale: Natali Anderson, "Marine Biologists Identify New Species of Beaked Whale," *Science News*, October 27, 2021, http://www.sci-news.com/biology/ramaris-beaked-whale-mesoplodon-eueu-10210.html.

17 Rice's whales: Patricia E. Rosel, Lynsey A. Wilcox, Tadasu K. Yamada, and Keith D. Mullin, "A New Species of Baleen Whale (*Balaenoptera*) from the Gulf of Mexico, with a Review of Its Geographic Distribution," *Marine Mammal Science* 37, no. 2 (2021): 577–610.

17 two new populations of pygmy blue whale: Sherry Landow, "New Population of Pygmy Blue Whales Discovered with Help of Bomb Detectors," *ScienceDaily*, June 8, 2021, https://www.sciencedaily.com/releases/2021/06/210608113226.htm.

Chapter 2. A Song in the Ocean

19 "Only if you love something": Lidija Haas, "Barbara Kingsolver: 'It Feels as Though We're Living Through the End of the World,'" *Guardian*, October 8, 2018.

21 "lovely curves": Author interview with Dr. Roger Payne.

21 "I removed the cigar": Roger Payne, liner notes to *Songs of the Humpback Whale*, CRM Records SWR 11, 1970, LP.

22 "maybe the day will be better": Author interview with Dr. Roger Payne.

23 "Tears flowed from our cheeks": Bill McQuay and Christopher Joyce, "It Took a Musician's Ear to Decode the Complex Song in Whale Calls," NPR, August 6, 2015.

23 more than seventy thousand whales killed each year: Robert C. Rocha, Jr., Phillip J. Clapham, and Yulia V. Ivashchenko, "Emptying the Oceans: A Summary of Industrial Whaling Catches in the 20th Century," *Marine Fisheries Review* 76, no. 4 (2015): 37–48.

23 "go save the whales": Ibid.

24 "series of sounds a 'song'": Roger S. Payne and Scott McVay, "Songs of Humpback Whales: Humpbacks Emit Sounds in Long, Predictable Patterns Ranging over Frequencies Audible to Humans," *Science* 173, no. 3997 (1971): 585–597, https://doi.org/10.1126/science.173.3997.585.

24 "quickly tuck their breaths in": Author interview with Dr. Roger Payne.

25 whales even employ rhyme: "It Took a Musician's Ear to Decode the Complex Song in Whale Calls."

25 songs evolve continuously: Ellen C. Garland, Luke Rendell, Luca Lamoni et al., "Song Hybridization Events During Revolutionary Song Change Provide Insights into Cultural Transmission in Humpback Whales," *Proceedings of the National Academy of Sciences of the United States of America* 114, no. 30 (2017): 7822–7829.

25 Katy Payne quoted Edward Sapir: Katy Payne and Ann Warde, "Humpback Whales: Composers of the Sea [Video]," Cornell Lab of Ornithology, All About Birds, May 21, 2014, https://www.allaboutbirds.org/news/humpback-whales-composers-of-the-sea-video/.

25 "Language moves down time in a current": Edward Sapir, *Language: An Introduction to the Study of Speech* (San Diego: Harcourt Brace, 2008), 1–4, 11, 150, 192, 218.

26 sing songs that have been compared to jazz: *Washington Post*, "The Jazz-like Sounds of Bowhead Whales," YouTube, video, April 4, 2018, https://www.youtube.com/watch?v=0GanRdxW7Fs.

26 363,661 whales were butchered in the 1970s: "Emptying the Oceans."

26 "capture the fancy of humanity": Invisibilia, "Two Heartbeats a Minute," Apple Podcasts, April 2020, https://podcasts.apple.com/us/podcast/two-heartbeats-a-minute/id953290300?i=1000467622321.

27 "and when they heard them they were stunned": Roger Payne, interview by Library of Congress, transcript, March 31, 2017, https://www.loc.gov/static/programs/national-recording-preservation-board/documents/RogerPayneInterview.pdf.

29 half of all the vertebrate animals have disappeared: Monique Grooten and Rosamunde E. A. Almond, eds., *Living Planet Report 2018: Aiming Higher* (Gland, Switzerland: WWF, 2018).

29 half of plants: Damian Carrington, "Humans Just 0.01% of All Life but Have Destroyed 83% of Wild Mammals—Study," *Guardian*, May 21, 2018.

29 "They make a desolation and call it peace": Quoted in Cornelius Tacitus, *Tacitus: Agricola*, ed. A. J. Woodman with C. S. Kraus (Cambridge, UK: Cambridge University Press, 2014).

29 25 billion farmed chickens: Tom Phillips, "How Many Birds Are Chickens?" Full Fact, February 27, 2020, https://fullfact.org/environment/how-many-birds-are-chickens/.

29 more plastic in them than fish: World Economic Forum, Ellen MacArthur Foundation, and McKinsey & Company, "The New Plastics Economy: Rethinking the Future of Plastics," Ellen MacArthur Foundation, 2016, https://ellenmacarthurfoundation.org/the-new-plastics-economy-rethinking-the-future-of-plastics.

30 three million whales: Daniel Cressey, "World's Whaling Slaughter Tallied," *Nature* 519, no. 7542 (2015): 140.

30 "We do not know the true nature of the entity we are destroying": Arthur C. Clarke, *Profiles of the Future: An Inquiry into the Limits of the Possible*, Millennium ed. (London: Phoenix Press, 2000).

31 prewhaling levels: Dr. Kirsten Thompson, "Humpback Whales Have Made a Remarkable Recovery, Giving Us Hope for the Planet," *Time*, May 16, 2020, https://time.com/5837350/humpback-whales-recovery-hope-planet/.

31 Reports in 2019: Alexandre N. Zerbini, Grant Adams, John Best et al., "Assessing the Recovery of an Antarctic Predator from Historical Exploitation," *Royal Society Open Science* 6, no. 10 (2019): 190368.

31 whales "rediscovering" the islands: British Antarctic Survey, "Blue Whales Return to Sub-Antarctic Island of South Georgia After Near Local Extinction," *ScienceDaily*, November 19, 2020, https://www.sciencedaily.com/releases/2020/11/201119103058.htm.

32 included Roger's whale songs: *The Golden Record. Greetings and Sounds of the Earth*, NASA Voyager Golden Record, NetFilmMusic, 2013, Track 3, 1:13, Spotify:track:5SnnD9Eac06j4O6TqBr3s2.

32 in its death throes engulfs Earth: K.-P. Schröder and Robert Connon Smith, "Distant Future of the Sun and Earth Revisited," *Monthly Notices of the Royal Astronomical Society* 386, no. 1 (2008): 155–163, https://doi.org/10.1111/j.1365-2966.2008.13022.x.

Chapter 3. The Law of the Tongue

35 "Imagine the possibilities": Robin Wall Kimmerer, *Braiding Sweetgrass: Indigenous Wisdom, Scientific Knowledge and the Teachings of Plants* (London: Penguin Books, 2020), 58.

36 These are called *symbioses*: Jennifer M. Lang and M. Eric Benbow, "Species Interactions and Competition," *Nature Education Knowledge* 4, no. 4 (2013): 8.

36 over a dozen fish swam out of its anus: Ed Yong, "How This Fish Survives in a Sea Cucumber's Bum," *National Geographic*, May 10, 2016, https://www.nationalgeographic.com/science/article/how-this-fish-survives-in-a-sea-cucumbers-bum.

37 "anal teeth": Dr. Chris Mah, "When Fish Live in Your Cloaca & How Anal Teeth Are Important!! The Pearlfish–Sea Cucumber Relationship!" *The Echinoblog* (blog), May 11, 2010, http://echinoblog.blogspot.com/2010/05/when-fish-live-in-your-cloaca-how-anal.html.

37 almost half a ton of barnacles: Mara Grunbaum, "What Whale Barnacles Know," *Hakai Magazine*, November 9, 2021, https://hakaimagazine.com/features/what-whale-barnacles-know/.

37 described the smell as "suffocating": Jonathan Kingdon, *East African Mammals: An Atlas of Evolution in Africa* (Chicago: University of Chicago Press, 1988), 89.

37 "either flee or become moribund": Ibid.

37 pistol shrimp team up with vertebrate goby fish: J. Lynn Preston, "Communication Systems and Social Interactions in a Goby-Shrimp Symbiosis," *Animal Behaviour* 26 (1978): 791–802.

38 old walls or gravestones: David Hill, "The Succession of Lichens on Gravestones: A Preliminary Investigation," *Cryptogamic Botany* 4 (1994): 179–186.

38 ideal homes for certain ants: Derek Madden and Truman P. Young, "Symbiotic Ants as an Alternative Defense Against Giraffe Herbivory in Spinescent *Acacia drepanolobium*," *Oecologia* 91, no. 2 (1992): 235–238.

39 resting on its goby pal's tail: Sam Ramirez and Jaclyn Calkins, "Symbiosis in Goby Fish and Alpheus Shrimp," Reed College, 2014, https://www.reed.edu/biology/courses/BIO342/2015_syllabus/2014_WEBSITES/sr_jc_website%202/index.html.

39 horses can sense the heart rates: Linda J. Keeling, Liv Jonare, and Lovisa Lanneborn, "Investigating Horse–Human Interactions: The Effect of a Nervous Human," *Veterinary Journal* 181, no. 1 (2009): 70–71.

39 "Mum, the police!": Tom Phillips, "Police Seize 'Super Obedient' Lookout Parrot Trained by Brazilian Drug Dealers," *Guardian*, April 24, 2019.

40 his friend Jack the chacma baboon: Simon Conway Morris, *Life's Solution: Inevitable Humans in a Lonely Universe* (Cambridge, UK: Cambridge University Press, 2003), 242.

40 Jack to be his apprentice signalman: "A Unique Signalman," *Railway Signal: Or, Lights Along the Line*, vol. 8 (London: The Railway Mission, 1890), 185.

41 distressed to see a monkey: Dorothy L. Cheney and Robert M. Seyfarth, *Baboon Metaphysics: The Evolution of a Social Mind* (Chicago: University of Chicago Press, 2007), 31.

42 persisted at least to the 1980s: Ibid., 33.

43 perhaps just ten left: Victor R. Rodríguez, "Will Exporting Farmed Totoaba Fix the Big Mess Pushing the World's Most Endangered Porpoise to Extinction?" *Hakai Magazine*, February 22, 2022, https://hakaimagazine.com/features/will-exporting-farmed-totoaba-fix-the-big-mess-pushing-the-worlds-most-endangered-porpoise-to-extinction/.

44 well over forty thousand years: Fran Dorey, "When Did Modern Humans Get to Australia?" Australian Museum, December 9, 2021, https://australian.museum/learn/science/human-evolution/the-spread-of-people-to-australia/.

45 as they were ten thousand years ago: John Upton, "Ancient Sea Rise Tale Told Accurately for 10,000 Years," *Scientific American*, January 26, 2015.

45 Among the Aboriginal people: "Whaling in Eden," Eden Community Access Centre, https://eden.nsw.au/whaling-in-eden. Excellent links to primary sources are also gathered here.

45 black-and-white-patterned ceremonial dress of warriors: "'King of Killers' Dead Body Washed Ashore: Whalers Ally for 100 Years," *Sydney Morning Herald*, September 18, 1930, 9.

45 crawl into the body of a dead whale: Fred Cahir, Ian Clark, and Philip Clarke, *Aboriginal Biocultural Knowledge in South-Eastern Australia: Perspectives of Early Colonists* (Collingwood, Victoria: CSIRO Publishing, 2018), 91.

45 perhaps for thousands of years: "Eden Killer Whale Museum: Old Tom's Skeleton," Bega Shire's Hidden Heritage, https://hiddenheritage.com.au/heritage-object/?object_id=8.

45 "Jaanda": "Becoming Beowa," Bundian Way, https://bundianway.com.au/becoming-beowa/.

45 killer whales became known as "beowa": *Aboriginal Biocultural Knowledge*, 90.

45 "departed ancestors": Danielle Clode, "Cooperative Killers Helped Hunt Whales," *Afloat*, December 2011, 3.

45 reward the killers: Clode, *Killers in Eden: The True Story of Killer Whales and Their Remarkable Partnership with the Whalers of Twofold Bay* (Crows Nest, NSW: Allen and Unwin, 2002).

45 Scottish whalers, the Davidsons: Killers of Eden, http://web.archive.org/web/*/www.killersofeden.com/. There are a family tree and extensive resources on this community website, which is no longer online but still fully accessible via the Wayback Machine.

46 Stranger, Skinner, and Jimmy: *Killers in Eden.*

46 ninety-three years old: "Meet the Whales of L-Pod from the Southern Resident Orca Population!" *Captain's Blog*, Orca Spirit Adventures, March 4, 2019, updated September 2020, https://orcaspirit.com/the-captains-blog/meet-the-whales-of-l-pod-in-2019-from-the-southern-resident-killer-whale-population/.

46 when the pod Old Tom belonged to: "King of the Killers", *Sydney Morning Herald*, September 18, 1930, https://www.smh.com.au/environment/conservation/the-king-of-the-killers-20100916-15er7.html.

46 slapping their tails on the surface: *Killers in Eden*, directed by Greg McKee, Australian Broadcasting Corporation, Vimeo, video, 2004, https://vimeo.com/47822835.

46 "the killers would let them know": *Killers in Eden.*

47 "Law of the Tongue": Bill Brown, "The Aboriginal Whalers of Eden," Australian Broadcast Corporation Local, audio, July 4, 2014, https://www.abc.net.au/local/audio/2013/10/29/3879462.htm.

47 found his body: "Eden Killer Whale Museum: Old Tom's Skeleton."

48 "such a combination": *Killers in Eden.*

48 reappeared in 1923: Blake Foden, "Old Tom: Anniversary of the Death of a Legend," *Eden Magnet*, September 16, 2014, https://www.edenmagnet.com.au/story/2563131/old-tom-anniversary-of-the-death-of-a-legend/.

48 "Oh God, what have I done?": "The King of the Killers," *Hawkesbury Gazette*, September 17, 2010.

49 U.S. Navy manuals: *U.S. Navy Diving Manual*, 1973. NAVSHIPS 0994-001-9010 (Washington, DC: Navy Department, 1973).

49 whistles of the individual dolphins: Elizabeth Preston, "Dolphins That Work with Humans to Catch Fish Have Unique Accent," *New Scientist*, October 2, 2017.

50 "gifts" of sea sponges: Giovanni Torre, "Dolphins Lavish Humans with Gifts During Lockdown on Australia's Cooloola Coast," *Telegraph*, May 21, 2020, https://www.telegraph.co.uk/news/2020/05/21/dolphins-lavish-humans-gifts-lockdown-australias-cooloola-coast/.

50 Pilot whales are attracted: Charlotte Curé, Ricardo Antunes, Filipa Samarra et al., "Pilot Whales Attracted to Killer Whale Sounds: Acoustically-Mediated Interspecific Interactions in Cetaceans," *PLoS One* 7, no. 12 (2012): e52201.

50 lost mother and calf pygmy sperm whale: Associated Press, "Dolphin Appears to Rescue Stranded Whales," NBC News, March 12, 2008, https://www.nbcnews.com/id/wbna23588063.

51 protect not just their fellow humpbacks: Robert L. Pitman et al., "Humpback Whales Interfering When Mammal-Eating Killer Whales Attack Other Species: Mobbing Behavior and Interspecific Altruism?" *Marine Mammal Science* 33, no. 1 (2017): 7–58, https://doi.org/10.1111/mms.12343.

51 days protecting the body: Jody Frediani, "Humpback Intervenes at Crime Scene, Returns Next Day with Friend," *Blog*, The Safina Center, January 20, 2021, https://www.safinacenter.org/blog/humpback-intervenes-at-crime-scene-returns-next-day-with-friend.

51 "Guboo" Ted Thomas: Brown, "The Aboriginal Whalers of Eden."

Chapter 4. The Joy of Whales

53 Joy was a fresh-faced grad student: Interview with Professor Joy Reidenberg, June 6, 2018.

56 chosen by Julius Caesar: Jason Daley, "Archeologists Discover Where Julius Caesar Landed in Britain," *Smithsonian Magazine*, November 30, 2017, https://www.smithsonianmag.com/smart-news/archaeologists-discover-where-julius-caesar-landed-britain-180967359/.

57 They can even explode in transit: MSNBC.com Staff, "Thar She Blows! Dead Whale Explodes," NBC News, January 29, 2004, https://www.nbcnews.com/id/wbna4096586.

57 rained giant chunks of blubber: Katie Shepherd, "Fifty Years Ago, Oregon Exploded a Whale in a Burst That 'Blasted Blubber Beyond All Believable Bounds,'" *Washington Post*, November 13, 2020, https://www.washingtonpost.com/nation/2020/11/13/oregon-whale-explosion-anniversary/.

59 "deep, heavy, agonizing groan": Herbert L. Aldrich, "Whaling," *Outing*, vol. 15, October 1899–March 1890, 113, Internet Archive, https://archive.org/details/outing15newy/page/n6/mode/1up.

59 hunting in the Antarctic: "Malcolm Clarke," obituary, *Telegraph*, July 30, 2013, https://www.telegraph.co.uk/news/obituaries/10211615/Malcolm-Clarke.html.

60 linked to Navy exercises: E. C. M. Parsons, "Impacts of Navy Sonar on Whales and Dolphins: Now Beyond a Smoking Gun?" *Frontiers in Marine Science* 4 (2017): 295, https://www.frontiersin.org/articles/10.3389/fmars.2017.00295/full.

60 literally scared to death: Mindy Weisberger, "Sonar Can Literally Scare Whales to Death, Study Finds," LiveScience, January 30, 2019, https://www.livescience.com/64635-sonar-beaked-whales-deaths.html.

61 sensitive to magnetic fields: Dorothee Kremers, Juliana López Marulanda, Martine Hausberger, and Alban Lemasson, "Behavioural Evidence of Magnetoreception in Dolphins: Detection of Experimental Magnetic Fields," *Naturwissenschaften* 101, no. 11 (2014): 907–911, https://doi.org/10.1007/s00114-014-1231-x.

63 "fundamental sensory and communication channel": Darlene R. Ketten, "The Marine Mammal Ear: Specializations for Aquatic Audition and Echolocation," in *The Evolutionary Biology of Hearing*, ed. Douglas B. Webster, Richard R. Fay, and Arthur N. Popper (New York: Springer-Verlag, 1992), 717–750.

63 thousands more of the receptive hairs: Ketten, "Structure and Function in Whale Ears," *Bioacoustics* 8, no. 1–2 (1997): 103–135.

63 *any machine we have yet built*: Sam H. Ridgway and Whitlow Au, "Hearing and Echolocation in Dolphins," *Encyclopedia of Neuroscience* 4 (2009): 1031–1039.

64 40 percent of its body length: "Sperm Whale," *Encyclopaedia Britannica*, updated March 30, 2021, https://www.britannica.com/animal/sperm-whale.

65 "one of the most noiseless": Thomas Beale, *The Natural History of the Sperm Whale: To Which Is Added a Sketch of a South-Sea Whaling Voyage, in Which the Author Was Personally Engaged* (London: J. Van Voorst, 1839).

66 louder than a jet engine: "Noise Sources and Their Effects," Purdue University Department of Chemistry, https://www.chem.purdue.edu/chemsafety/Training/PPETrain/dblevels.htm.

66 interrogate the ocean: Eduardo Mercado III, "The Sonar Model for Humpback Whale Song Revised," *Frontiers in Psychology* 9 (2018): 1156, https://doi.org/10.3389/fpsyg.2018.01156.

66 over seventy different coda types: Rendell and Hal Whitehead, "Vocal Clans in Sperm Whales (*Physeter macrocephalus*)," *Proceedings of the Royal Society B: Biological Sciences* 270, no. 1512 (2003): 225–231.

66 the glue that holds their cooperative lives together: Whitehead, *Sperm Whales: Social Evolution in the Ocean* (Chicago: University of Chicago Press, 2003).

67 "broadest range of acoustic channels": "The Marine Mammal Ear."

69 These leviathans operate nurseries: Shane Gero, Dan Engelhaupt, Rendell, and Whitehead, "Who Cares? Between-Group Variation in Alloparental Caregiving in Sperm Whales," *Behavioral Ecology* 20, no. 4 (2009): 838–843.

69 protected within this "marguerite": Pitman, Lisa T. Ballance, Sarah I. Mesnick, and Susan J. Chivers, "Killer Whale Predation on Sperm Whales: Observations and Implications," *Marine Mammal Science* 17, no. 3 (2001): 494–507, https://doi.org/10.1111/j.1748-7692.2001.tb01000.x.

69 providing food for those less able to hunt: Kerry Lotzof, "Life in the Pod: The Social Lives of Whales," Natural History Museum, https://www.nhm.ac.uk/discover/social-lives-of-whales.html.

69 dubbed these "vocal clans": Rendell and Whitehead, "Vocal Clans in Sperm Whales."

70 whales have "cultures": Rendell and Whitehead, "Culture in Whales and Dolphins," *Behavioral and Brain Sciences* 24, no. 2 (2001): 309–324.

Chapter 5. "Some Sort of Stupid, Big Fish"

71 "atoms with consciousness": Richard P. Feynman, *The Pleasure of Finding Things Out: The Best Short Works of Richard Feynman*, ed. Jeffrey Robbins (New York: Basic Books, 1999), 144.

73 scan the brains of living dolphins: Ridgway, Dorian Houser, James Finneran et al., "Functional Imaging of Dolphin Brain Metabolism and Blood Flow," *Journal of Experimental Biology* 209 (Pt. 15) (2006): 2902–2910.

77 miles behind the sperm whale: Mind Matters, "Are Whales Smarter Than We Are?" *News Blog, Scientific American*, January 15, 2008, https://blogs.scientificamerican.com/news-blog/are-whales-smarter-than-we-are/.

77 10.5 billion cerebral neurons: Ursula Dicke and Gerhard Roth, "Neuronal Factors Determining High Intelligence," *Philosophical Transactions of the Royal Society B: Biological Sciences* 371, no. 1685 (2016): 20150180.

77 important for cognition, too: R. Douglas Fields, "The Other Half of the Brain," *Scientific American*, April 2004, https://www.scientificamerican.com/article/the-other-half-of-the-brain/.

78 One study concluded that humans: Dicke and Roth, "Neuronal Factors."

78 "We suck at being able": David Grimm, "Are Dolphins Too Smart for Captivity?" *Science* 332, no. 6029 (2011): 526–529, https://doi.org/10.1126/science.332.6029.526.

79 "the smartest bears and the dumbest tourists": Lynn Smith, "My Take: Dumb and Dumber," *Holland Sentinel*, November 20, 2020, https://www.hollandsentinel.com/story/opinion/columns/2020/11/20/my-take-dumb-and-dumber/114997362/.

80 "the same way in whales": Author's interview with Professor Patrick R. Hof, June 8, 2018.

80 brains of neuroscientists fizzing: Patrick R. Hof and Estel Van der Gucht, "Structure of the Cerebral Cortex of the Humpback Whale, *Megaptera novaeangliae* (*Cetacea, Mysticeti, Balaenopteridae*)," *Anatomical Record* 290, no. 1 (Hoboken, NJ, 2007): 1–31.

80 like lemurs: Esther A. Nimchinsky, Emmanuel Gilissen, John M. Allman et al., "A Neuronal Morphologic Type Unique to Humans and Great Apes," *Proceedings of the National Academy of Sciences of the United States of America* 96, no. 9 (1999): 5268–5273.

80 We are distant relatives: Maureen A. O'Leary, Jonathan I. Bloch, John J. Flynn et al., "The Placental Mammal Ancestor and the Post–K-Pg Radiation of Placentals," *Science* 339, no. 6120 (2013): 662–667.

81 "'express trains' of the nervous system": Andy Coghlan, "Whales Boast the Brain Cells That 'Make Us Human,'" *New Scientist*, November 27, 2006, https://www.newscientist.com/article/dn10661-whales-boast-the-brain-cells-that-make-us-human/.

81 brains of cows, sheep, deer, horses: Mary Ann Raghanti, Linda B. Spurlock, F. Robert Treichler et al., "An Analysis of von Economo Neurons in the Cerebral Cortex of Cetaceans, Artiodactyls, and Perissodactyls," *Brain Structure & Function* 220, no. 4 (2015): 2303–2314, https://doi.org/10.1007/s00429-014-0792-y.

82 "We don't even understand the brain of a worm": Rachel Tompa, "5 Unsolved Mysteries About The Brain," Allen Institute, March 14, 2019, https://alleninstitute.org/what-we-do/brain-science/news-press/articles/5-unsolved-mysteries-about-brain.

82 In 2007, Lori Marino: Lori Marino, Richard C. Connor, R. Ewan Fordyce et al., "Cetaceans Have Complex Brains for Complex Cognition," *PLoS Biology* 5, no. 5 (2007): e139.

84 The brain of a bottlenose dolphin (center) flanked by that of a wild pig (left) and a human (right): G. G. Mascetti, "Unihemispheric Sleep and Asymmetrical Sleep: Behavioral, Neurophysiological, and Functional Perspectives," *Nature and Science of Sleep*, vol. 8 (2016): 221–238.

85 He told me how he had been "scanned": Author's interview with Duncan Brake, November 20, 2019, Turks and Caicos.

Chapter 6. The Search for Animal Language

87 like a "momentary squabble": *Deep Voices: The Second Whale Record*, Capitol Records ST-11598, 1977, LP.

89 with very different answers: Ewa Dąbrowska, "What Exactly Is Universal Grammar, and Has Anyone Seen It?" *Frontiers in Psychology* 6 (2015): 852, https://doi.org/10.3389/fpsyg.2015.00852.

89 *We are born a blank slate*: B. F. Skinner, *Verbal Behavior* (New York: Appleton-Century-Crofts, 1957).

89 *We are born with a special human Universal Grammar!*: Noam Chomsky, *Knowledge of Language: Its Nature, Origin and Use* (New York: Praeger, 1986).

89 *A language instinct!*: Steven Pinker, *The Language Instinct: How the Mind Creates Language* (London: Penguin Books, 2003).

89 *There is no Universal Grammar!*: Philip Lieberman, *Human Language and Our Reptilian Brain: The Subcortical Bases of Speech, Syntax, and Thought* (Cambridge, MA: Harvard University Press, 2000).

89 *But humans can build language from our cultures!*: Daniel Everett, *Don't Sleep, There Are Snakes!* (London: Profile Books, 2009), 243.

89 *There is no "seat" of language in the brain*: Lieberman, "Human Language and Our Reptilian Brain: The Subcortical Bases of Speech, Syntax, and Thought," *Perspectives in Biology and Medicine* 44 (2001): 32–51.

89 ***Recursiveness is what makes our language special!*****:** Marc D. Hauser, Chomsky, and W. Tecumseh Fitch, "The Faculty of Language: What Is It, Who Has It, and How Did It Evolve?" *Science* 298, no. 5598 (November 22, 2002): 1569–1579.

89 ***True language is only possible verbally*****:** John L. Locke and Barry Bogin, "Language and Life History: A New Perspective on the Development and Evolution of Human Language," *Behavioural and Brain Sciences* 29, no. 3 (2006): 259–325.

89 ***Language is a multifaceted phenomenon*****:** Sławomir Wacewicz and Przemysław Żywiczyński, "Language Evolution: Why Hockett's Design Features are a Non-Starter," *Biosemiotics* 8, no. 1 (2015): 29–46.

89 **to qualify as a language:** Edmund West, "William Stokoe—American Sign Language scholar," *British Deaf News,* January 30, 2020, https://www.britishdeafnews.co.uk/william-stokoe/.

90 **"waving a red cape in front of a bull":** Con Slobodchikoff, *Chasing Doctor Dolittle: Learning the Language of Animals* (New York: St. Martin's Press, 2012).

90 **"the one historical constant":** Frans de Waal, "The Brains of the Animal Kingdom," *Wall Street Journal,* March 22, 2013, https://www.wsj.com/articles/SB10001424127887323869604578370574285382756.

90 **thought uniquely human:** Kate Douglas, "Six 'Uniquely Human' Traits Now Found in Animals," *New Scientist,* May 22, 2008, https://www.newscientist.com/article/dn13860-six-uniquely-human-traits-now-found-in-animals/.

91 **multimodal:** James P. Higham and Eileen A. Hebets, "An Introduction to Multimodal Communication," *Behavioral Ecology and Sociobiology* 67, no. 9 (2013): 1381–1388.

91 **fifteen glands:** Laura Bortolotti and Cecilia Costa, "Chemical Communication in the Honey Bee Society," in *Neurobiology of Chemical Communication*, ed. Carla Mucignat-Caretta (Boca Raton, FL: Taylor & Francis, 2014).

91 **winged semaphore:** Meredith C. Miles and Matthew J. Fuxjager, "Synergistic Selection Regimens Drive the Evolution of Display Complexity in Birds of Paradise," *Journal of Animal Ecology* 87, no. 4 (2018): 1149–1159.

91 **reflectiveness of our skin in milliseconds:** Alejandra López Galán, Wen-Sung Chung, and N. Justin Marshall, "Dynamic Courtship Signals and Mate Preferences in *Sepia plangon*," *Frontiers in Physiology* 11 (2020): 845.

91 **Cows have twice the hearing range we do:** Richard E. Berg, "Infrasonics," *Encyclopaedia Britannica*, https://www.britannica.com/science/infrasonics.

91 **they become excited—such as when tickled:** Ashwini J. Parsana, Nanxin Li, and Thomas H. Brown, "Positive and Negative Ultrasonic Social Signals Elicit Opposing Firing Patterns in Rat Amygdala," *Behavioural Brain Research* 226, no. 1 (2012): 77–86.

92 **published a linguistics textbook:** Charles F. Hockett, *A Course in Modern Linguistics* (New York: Macmillan, 1958), section 64, 569–586.

93 **"design features" of language:** Hockett, "The Origin of Speech," *Scientific American* 203, no. 3 (1960): 88–97.

93 **some "very bitter disputes":** Guy Cook, *Applied Linguistics* (Oxford: Oxford University Press, 2003).

95 could do in real time: Bart de Boer, Neil Mathur, and Asif A. Ghazanfar, "Monkey Vocal Tracts Are Speech-Ready," *Science Advances* 2, no. 12 (2016): e1600723.

96 somewhat whispery: Michael Price, "Why Monkeys Can't Talk—and What They Would Sound Like If They Could," *Science*, December 9, 2016, https://www.sciencemag.org/news/2016/12/why-monkeys-can-t-talk-and-what-they-would-sound-if-they-could.

96 is contested (sometimes fiercely): Pedro Tiago Martins and Cedric Boeckx, "Vocal Learning: Beyond the Continuum," *PLoS Biology* 18, no. 3 (2020): e3000672.

96 The vast majority of mammals and many of the birds: Andreas Nieder and Richard Mooney, "The Neurobiology of Innate, Volitional and Learned Vocalizations in Mammals and Birds," *Philosophical Transactions of the Royal Society B: Biological Sciences* 375, no. 1789 (2020): 20190054.

97 "You bloody fool!": Ben Panko, "Listen to Ripper the Duck Say 'You Bloody Fool!' " *Smithsonian Magazine*, September 9, 2021, https://www.smithsonianmag.com/smart-news/listen-ripper-duck-say-you-bloody-fool-180978613/.

97 blow out Korean words: Russell Goldman, "Korean Words, Straight from the Elephant's Mouth," *New York Times*, May 26, 2016, https://www.nytimes.com/2016/05/27/world/what-in-the-world/korean-words-straight-from-the-elephants-mouth.html.

97 gruff New England accent: New England Aquarium, "Hoover the Talking Seal," YouTube, video, November 28, 2007, https://www.youtube.com/watch?v=prrMaLrkc5U&t=8s.

97 "so aggressively nuanced": Roger Payne, email, January 2022.

97 sing two different songs at the same time: Tobias Riede and Franz Goller, "Functional Morphology of the Sound-Generating Labia in the Syrinx of Two Songbird Species," *Journal of Anatomy* 216, no. 1 (2010): 23–36.

97 beluga whale called Noc: Ewen Callaway, "The Whale That Talked," *Nature*, 2012, https://doi.org/10.1038/nature.2012.11635.

98 "wanted to make a connection": Charles Siebert, "The Story of One Whale Who Tried to Bridge the Linguistic Divide Between Animals and Humans," *Smithsonian Magazine*, June 2014, https://www.smithsonianmag.com/science-nature/story-one-whale-who-tried-bridge-linguistic-divide-between-animals-humans-180951437/.

98 derived from American Sign Language: R. Allen Gardner and Beatrice T. Gardner, "Teaching Sign Language to a Chimpanzee," *Science* 165, no. 3894 (1969): 664–672.

98 ersatz symbolic languages: David Premack, "On the Assessment of Language Competence in the Chimpanzee," in *Behavior of Nonhuman Primates*, vol. 4, ed. Allan M. Schrier and Fred Stollnitz (New York: Academic Press, 1971), 186–228.

98 screens of computer interfaces: Duane M. Rumbaugh, Timothy V. Gill, Josephine V. Brown et al., "A Computer-Controlled Language Training System for

Investigating the Language Skills of Young Apes," *Behavior Research Methods & Instrumentation* 5, no. 5 (1973): 385–392.

99 The chimpanzees of Budongo Forest: Raphaela Heesen et al., "Linguistic Laws in Chimpanzee Gestural Communication," *Proceedings of the Royal Society B: Biological Sciences* 286, no. 1896 (2019), https://doi.org/10.1098/rspb.2018.2900.

100 two-year-old humans use: Verena Kersken et al, "A gestural repertoire of 1- to 2-year-old human children: in search of the ape gestures," *Animal Cognition* 22 (2019): 577–595.

100 By the age of two: Email to author, April 2022.

100 he stopped asking: Steven M. Wise, *Drawing the Line* (Cambridge, MA: Perseus Books, 2002), 107.

100 "a kind of 'Dr. Dolittle' moment": Irene M. Pepperberg, "Animal Language Studies: What Happened?" *Psychonomic Bulletin & Review* 24 (2017): 181–185, https://doi.org/10.3758/s13423-016-1101-y.

101 "whatever it was that apes didn't have": Roger S. Fouts, "Language: Origins, Definitions and Chimpanzees," *Journal of Human Evolution* 3, no. 6 (1974): 475–482.

101 80 percent of the time: "Ask the Scientists: Irene Pepperberg," Scientific American Frontiers Archives, PBS, Internet Archive Wayback Machine, https://web.archive.org/web/20071018070320/http:/www.pbs.org/safarchive/3_ask/archive/qna/3293_pepperberg.html.

102 "bad, sad, bad, frown, cry-frown, sad": Thori, "Koko the Gorilla Cries over the Loss of a Kitten," YouTube, video, December 8, 2011, https://www.youtube.com/watch?v=CQCOHUXmEZg.

103 "literally surrounded with evidence": Slobodchikoff, *Chasing Doctor Dolittle.*

104 Vervet monkeys also: Seyfarth, Cheney, and Peter Marler, "Vervet Monkey Alarm Calls: Semantic Communication in a Free-Ranging Primate," *Animal Behaviour* 28, no. 4 (1980): 1070–1094.

104 Much evidence has now been gathered: Klaus Zuberbühler, "Survivor Signals: The Biology and Psychology of Animal Alarm Calling," *Advances in the Study of Behavior* 40 (2009): 277–322.

105 Jungle fowl seem: Nicholas E. Collias, "The Vocal Repertoire of the Red Junglefowl: A Spectrographic Classification and the Code of Communication," *Condor* 89, no. 3 (1987): 510–524.

105 caged chickens: Christopher S. Evans, Linda Evans, and Marler, "On the Meaning of Alarm Calls: Functional Reference in an Avian Vocal System," *Animal Behaviour* 46, no. 1 (1993): 23–38.

105 as do lemurs: Claudia Fichtel, "Reciprocal Recognition of Sifaka (*Propithecus verreauxi verreauxi*) and Redfronted Lemur (*Eulemur fulvus rufus*) Alarm Calls," *Animal Cognition* 7, no. 1 (2004): 45–52.

105 "Alan, Alan": BBC, "Alan!.. Alan!.. Steve! Walk on the Wild Side—BBC," YouTube, video, March 19, 2009, https://www.youtube.com/watch?v=xaPepcVepCg.

106 it seems prairie dogs have nouns: Slobodchikoff, Andrea Paseka, and Jennifer L. Verdolin, "Prairie Dog Alarm Calls Encode Labels About Predator Colors," *Animal Cognition* 12, no. 3 (2009): 435–439.

106 silhouettes of coyotes: Email to author, April 2022.

107 meaning is encoded in bird calls: Sabrina Engesser, Jennifer L. Holub, Louis G. O'Neill et al., "Chestnut-Crowned Babbler Calls Are Composed of Meaningless Shared Building Blocks," *Proceedings of the National Academy of Sciences of the United States of America* 116, no. 39 (2019): 19579–19584.

107 the pied babbler: Engesser et al., "Internal acoustic structuring in pied babbler recruitment cries specifies the form of recruitment," *Behavioral Ecology* 29, no. 5 (2018): 1021–1030.

107 other threat calls: Engesser et al., "Meaningful call combinations and compositional processing in the southern pied babbler." *Proceedings of the National Academy of Sciences* 113, no. 21 (2016): 5976–5981.

107 within a month: T. N. Suzuki et al., "Experimental evidence for compositional syntax in bird calls," *Nature Communication* 7 (2016): 10986.

108 "preponderance of evidence": Holly Root-Gutteridge, interview with author, September 1, 2019.

108 lamented that arguments: Slobodchikoff, *Chasing Doctor Dolittle.*

Chapter 7. Deep Minds: Cetacean Culture Club

110 "Never trust a species": Terry Pratchett, *Pyramids* (London: Corgi, 2012), 207.

111 use keyboards to communicate: Harvest Books, "The Dolphin in the Mirror: Keyboards," YouTube, video, July 8, 2011, https://www.youtube.com/watch?v=3IqRPaAYm4I.

111 a sponge to protect their rostrums: Virginia Morell, "Why Dolphins Wear Sponges," *Science*, July 20, 2011, https://www.science.org/content/article/why-dolphins-wear-sponges.

111 lured seagulls into his pool: Bjorn Carey, "How Killer Whales Trap Gullible Gulls," NBC News, February 3, 2006, https://www.nbcnews.com/id/wbna11163990.

111 seaweed just out of their grasp: Joe Noonan, "Wild Dolphins Playing w/Seaweed & Snorkeler: Slomo—Very Touching," YouTube, video, April 28, 2017, https://www.youtube.com/watch?v=5_DLhtq5Ctg.

111 bow-ride blue whales: Capt. Dave's Dana Point Dolphin & Whale Watching Safari, "Dolphins 'Bow Riding' with Blue Whales off Dana Point," YouTube, video, July 27, 2012, https://www.youtube.com/watch?v=wfEdki3LwUY.

111 surf and leap from breaking waves: BBC, "Glorious Dolphins Surf the Waves Just for Fun: Planet Earth: A Celebration—BBC," YouTube, video, September 1, 2020, https://www.youtube.com/watch?v=6HRMHejDHHm.

112 circle and toy with human swimmers: Dylan Brayshaw, "Orcas Approaching Swimmer FULL VERSION (Unedited)," YouTube, video, December 16, 2019, https://www.youtube.com/watch?v=gVmieqjU0E8.

112 frolic with kayakers: *Wall Street Journal*, "Orca and Kayaker Encounter Caught on Drone Video," YouTube, video, September 9, 2016, https://www.youtube.com/watch?v=eoUVufAuEw0.

112 hula hoop around their pool: Stan A. Kuczaj II and Rachel T. Walker, "Dolphin Problem Solving," in *The Oxford Handbook of Comparative Cognition*, ed. Thomas R.

Zentall and Edward A. Wasserman (New York: Oxford University Press, 2012), 736–756.

112 perfect bubble rings: Brenda McCowan, Marino, Erik Vance et al., "Bubble Ring Play of Bottlenose Dolphins (*Tursiops truncatus*): Implications for Cognition," *Journal of Comparative Psychology* 114, no. 1 (2000): 98.

112 seem to comprehend pointing: Adam A. Pack and Louis M. Herman, "Bottlenosed Dolphins (*Tursiops truncatus*) Comprehend the Referent of Both Static and Dynamic Human Gazing and Pointing in an Object-Choice Task," *Journal of Comparative Psychology* 118, no. 2 (2004): 160.

112 "an animal that has no arms, hands, fingers": Justin Gregg, *Are Dolphins Really Smart? The Mammal Behind the Myth* (Oxford: Oxford University Press, 2013).

113 point things out to their trainers: Mark J. Xitco, John D. Gory, and Kuczaj, "Spontaneous Pointing by Bottlenose Dolphins (*Tursiops truncatus*)," *Animal Cognition* 4, no. 2 (2001): 115–123.

113 dead dolphins to one another: K. M. Dudzinski, M. Saki, K. Masaki et al., "Behavioural Observations of Bottlenose Dolphins Towards Two Dead Conspecifics," *Aquatic Mammals* 29, no. 1 (2003): 108–116.

113 somewhat like a name: Morell, "Dolphins Can Call Each Other, Not by Name, but by Whistle," *Science*, February 20, 2013, https://www.science.org/content/article/dolphins-can-call-each-other-not-name-whistle.

113 copy each other's whistles: Stephanie L. King, Heidi E. Harley, and Vincent M. Janik, "The Role of Signature Whistle Matching in Bottlenose Dolphins, *Tursiops truncatus*," *Animal Behaviour* 96 (2014): 79–86.

113 remember their friends' signature whistles for more than two decades: Jason N. Bruck, "Decades-Long Social Memory in Bottlenose Dolphins," *Proceedings of the Royal Society B: Biological Sciences* 280, no. 1768 (2013): 20131726.

113 impressions of humpback whale: Mary Bates, "Dolphins Speaking Whale?" American Association for the Advancement of Science, February 6, 2012, https://www.aaas.org/dolphins-speaking-whale.

113 mimic each other's sounds when they fight: Laura J. May-Collado, "Changes in Whistle Structure of Two Dolphin Species During Interspecific Associations," *Ethology* 116, no. 11 (2010): 1065–1074.

113 Killer whales have been observed imitating other whales: John K. B. Ford, "Vocal Traditions Among Resident Killer Whales (*Orcinus orca*) in Coastal Waters of British Columbia," *Canadian Journal of Zoology* 69, no. 6 (1991): 1454–1483.

113 barking sounds like sea lions: Andrew D. Foote, Rachael M. Griffin, David Howitt et al., "Killer Whales Are Capable of Vocal Learning," *Biology Letters* 2, no. 4 (2006): 509–512.

115 study of dolphins and their "language": Christopher Riley, "The Dolphin Who Loved Me: The NASA-Funded Project That Went Wrong," *Guardian*, June 8, 2014, https://www.theguardian.com/environment/2014/jun/08/the-dolphin-who-loved-me.

116 "weird ideas about dolphins": Gregg, *Are Dolphins Really Smart?*

116 "rock-corn": Sy Montgomery, *Birdology: Adventures with a Pack of Hens, a Peck of Pigeons, Cantankerous Crows, Fierce Falcons, Hip Hop Parrots, Baby Hummingbirds, and One Murderously Big Living Dinosaur* (Riverside, CA: Atria Books, 2010), 197.

116 the chimp signed "water-bird": Benedict Carey, "Washoe, a Chimp of Many Words, Dies at 42," *New York Times*, November 1, 2007, https://www.nytimes.com/2007/11/01/science/01chimp.html.

117 given an underwater telephone: Crispin Boyer, "Secret Language of Dolphins," *National Geographic Kids*, https://kids.nationalgeographic.com/nature/article/secret-language-of-dolphins.

117 complex communication systems: Herman, Sheila L. Abichandani, Ali N. Elhajj et al., "Dolphins (*Tursiops truncatus*) Comprehend the Referential Character of the Human Pointing Gesture," *Journal of Comparative Psychology* 113, no. 4 (1999): 347.

118 "intentionally wrong symbol sequences": Gregg, *Are Dolphins Really Smart?*

118 classify objects based on their shape: Kelly Jaakkola, Wendi Fellner, Linda Erb et al., "Understanding of the Concept of Numerically 'Less' by Bottlenose Dolphins (*Tursiops truncatus*)," *Journal of Comparative Psychology* 119, no. 3 (2005): 296.

118 number: Annette Kilian, Sevgi Yaman, Lorenzo von Fersen, and Onur Güntürkün, "A Bottlenose Dolphin Discriminates Visual Stimuli Differing in Numerosity," *Animal Learning & Behavior* 31, no. 2 (2003): 133–142.

118 and relative size: Mercado III, Deirdre A. Killebrew, Pack et al., "Generalization of 'Same–Different' Classification Abilities in Bottlenosed Dolphins," *Behavioural Processes* 50, no. 2–3 (2000): 79–94.

118 concept of "humans": Gregg, *Are Dolphins Really Smart?* 100.

119 supposing it to be a threat: Charles J. Meliska, Janice A. Meliska, and Harman V. S. Peeke, "Threat Displays and Combat Aggression in *Betta splendens* Following Visual Exposure to Conspecifics and One-Way Mirrors," *Behavioral and Neural Biology* 28, no. 4 (1980): 473–486.

120 showed it to some bottlenose dolphins: Diana Reiss and Marino, "Mirror Self-Recognition in the Bottlenose Dolphin: A Case of Cognitive Convergence," *Proceedings of the National Academy of Sciences of the United States of America* 98, no. 10 (2001): 5937–5942.

120 "sequential intromission attempts": Reiss, email to author, December 20, 2021.

121 Bayley was able to mirror self-recognize: Rachel Morrison and Reiss, "Precocious Development of Self-Awareness in Dolphins," *PLoS One* 13, no. 1 (2018): e0189813.

121 human children, who generally start: James Gorman, "Dolphins Show Self-Recognition Earlier Than Children," *New York Times*, January 10, 2018, https://www.nytimes.com/2018/01/10/science/dolphins-self-recognition.html.

121 was supported when their larger relatives: Fabienne Delfour and Ken Marten, "Mirror Image Processing in Three Marine Mammal Species: Killer Whales

(*Orcinus orca*), False Killer Whales (*Pseudorca crassidens*) and California Sea Lions (*Zalophus californianus*)," *Behavioural Processes* 53, no. 3 (2001): 181–190.

122 allow us to get a reflection: Carolyn Wilkie, "The Mirror Test Peers into the Workings of Animal Minds" *Scientist*, February 21, 2019, https://www.the-scientist.com/news-opinion/the-mirror-test-peers-into-the-workings-of-animal-minds-65497.

122 can conceive of having a body: Herman, "Body and Self in Dolphins," *Consciousness and Cognition* 21, no. 1 (2012): 526–545.

122 that bottlenose dolphins . . . : Herman, "Vocal, Social, and Self-Imitation by Bottlenosed Dolphins," in *Imitation in Animals and Artifacts*, ed. Kerstin Dautenhahn and Chrystopher L. Nehaniv (Cambridge, MA: MIT Press, 2002), 63–108.

122 and killer whales can choose what activity: José Z. Abramson, Victoria Hernández-Lloreda, Josep Call, and Fernando Colmenares, "Experimental Evidence for Action Imitation in Killer Whales (*Orcinus orca*)," *Animal Cognition* 16, no. 1 (2013): 11–22.

122 invent a new task to perform: Mercado, Scott O. Murray, Robert K. Uyeyama et al., "Memory for Recent Actions in the Bottlenosed Dolphin (*Tursiops truncatus*): Repetition of Arbitrary Behaviors Using an Abstract Rule," *Animal Learning & Behavior* 26, no. 2 (1998): 210–218.

123 other people should know about this: Reiss and Marino, "Mirror Self-Recognition in the Bottlenose Dolphin."

123 ethics of how we treat them: Grimm, "Are Dolphins Too Smart for Captivity?"

123 news reports enthralling the nation: Katherine Bishop, "Flotilla Drives Errant Whale into Salt Water," *New York Times*, November 4, 1985, https://www.nytimes.com/1985/11/04/us/flotilla-drives-errant-whale-into-salt-water.html.

126 Spectrogram of dolphin volcalizations: Eric A. Ramos and Diana Reiss, 2014, "Foraging-related calls produced by bottlenose dolphins." Paper presented at the 51st Annual Conference of the Animal Behaviour Society, Princeton NJ, Aug 9-14, 2014.

Chapter 8. The Sea Has Ears

128 *"Nānā ka maka"*: Mary Kawena Pukui, ed. *'Olelo No'eau: Hawaiian Proverbs & Poetical Sayings*, Bernice P. Bishop Museum special publication no. 71 (Honolulu: Bishop Museum Press, 1983).

128 "mammal brothers and sisters": Richard Brautigan, *All Watched Over by Machines of Loving Grace* (San Francisco: Communication Company, 1967).

129 they began with the Big Island: Christine Hitt, "The Sacred History of Maunakea," *Honolulu*, August 5, 2019, https://www.honolulumagazine.com/the-sacred-history-of-maunakea/.

130 had just died:. Christie Wilcox, "'Lonely George' the Snail Has Died, Marking the Extinction of His Species," *National Geographic*, January 9, 2019, https://www.nationalgeographic.co.uk/animals/2019/01/lonely-george-snail-has-died-marking-extinction-his-species.

131 11 species of birds were declared extinct: Brian Hires, "U.S. Fish and Wildlife Service Proposes Delisting 23 Species from Endangered Species Act Due to Extinction," press release, U.S. Fish and Wildlife Service (website), September 29, 2021, https://www.fws.gov/news/ShowNews.cfm?ref=u.s.-fish-and-wildlife-service-proposes-delisting-23-species-from-&_ID=37017.

138 birds he was observing had different accents: Kristina L. Paxton, Esther Sebastián-González, Justin M. Hite et al., "Loss of Cultural Song Diversity and the Convergence of Songs in a Declining Hawaiian Forest Bird Community," *Royal Society Open Science* 6, no. 8 (2019): 190719.

143 twelve thousand humpback whales: Anke Kügler, Marc O. Lammers, Eden J. Zang et al., "Fluctuations in Hawaii's Humpback Whale *Megaptera novaeangliae* Population Inferred from Male Song Chorusing off Maui," *Endangered Species Research* 43 (2020): 421–434, https://doi.org/10.3354/esr01080.

143 was called "the Blob": Eli Kintisch, "'The Blob' Invades Pacific, Flummoxing Climate Experts," *Science* 348, no. 6230 (April 3, 2015): 17–18, https://www.science.org/doi/10.1126/science.348.6230.17.

Chapter 9. Animalgorithms

145 "Machines take me by surprise with great frequency": A. M. Turing, "Computing Machinery and Intelligence," *Mind* (New Series) 59, no. 236 (1950): 433–460.

145 you could engrave sound: Thomas A. Edison, "The Talking Phonograph," *Scientific American* 37, no. 25 (1877): 384–385.

145 made a unique recording: Arthur A. Allen and Peter Paul Kellogg, "Song Sparrow," audio, Macaulay Library, The Cornell Lab of Ornithology, May 18, 1929, digitized December 12, 2001, https://macaulaylibrary.org/asset/16737.

145 search for ivory-billed woodpeckers: Chelsea Steinauer-Scudder, "The Lord God Bird: Apocalyptic Prophecy & the Vanishing of Avifauna," *Emergence Magazine*, July 1, 2020, https://emergencemagazine.org/essay/the-lord-god-bird/.

148 in "informal settings": International Bioacoustics Society (IBAC) website, https://www.ibac.info.

149 people were using sound recording and manipulation devices: Examples from IBAC presentations can be found at "Programme IBAC 2019," IBAC, https://2019.ibac.info/programme.

150 71 percent of species have female song: Katharina Riebel, Karan J. Odom, Naomi E. Langmore, and Michelle L. Hall, "New Insights from Female Bird Song: Towards an Integrated Approach to Studying Male and Female Communication Roles," *Biology Letters* 15, no. 4 (2019): 20190059, http://doi.org/10.1098/rsbl.2019.0059.

150 "There's been almost a century and a half of research": Bates, "Why Do Female Birds Sing?" *Animal Minds* (blog), *Psychology Today*, August 26, 2019, https://www.psychologytoday.com/gb/blog/animal-minds/201908/why-do-female-birds-sing.

151 Merlin Sheldrake talks about how helpful queer theory: Whitney Bauck, "Mythos and Mycology," *Atmos*, June 14, 2021, https://atmos.earth/fungi-mushrooms-merlin-sheldrake-interview/.

152 New Zealand bellbird: Wesley H. Webb, M. M. Roper, Matthew D. M. Pawley, Yukio Fukuzawa, A. M. T. Harmer, and D. H. Brunton, "Sexually distinct song cultures across a songbird metapopulation," *Frontiers in Ecology and Evolution*, 9, 2021, https://www.frontiersin.org/article/10.3389/fevo.2021.755633.

152 Koe had massively sped up his classifying... of the 21,500 song units: Fukuzawa, Webb, Pawley et al., "*Koe*: Web-Based Software to Classify Acoustic Units and Analyse Sequence Structure in Animal Vocalizations," *Methods in Ecology and Evolution* 11, no. 3 (2020): 431–441.

153 Across this flesh canvas is engraved: Steven K. Katona and Whitehead, "Identifying Humpback Whales Using Their Natural Markings," *Polar Record* 20, no. 128 (1981): 439–444.

154 $25,000 reward from Google: Ted Cheeseman, email to author, November 28, 2021.

155 poured into his "fully automated": Cheeseman et al., "Advanced Image Recognition: A Fully Automated, High-Accuracy Photo-Identification Matching System for Humpback Whales," *Mammalian Biology* (2021), http://doi.org/10.1007/s42991-021-00180-9.

156 Ted looked up its record: "Prime Suspect," Humpback Whale CRC-12564, Happywhale, https://happywhale.com/individual/1437.

158 male mice sing different songs: Jonathan Chabout, Abhra Sarkar, David B. Dunson, and Erich D. Jarvis, "Male Mice Song Syntax Depends on Social Contexts and Influences Female Preferences," *Frontiers in Behavioral Neuroscience* 9 (April 1, 2015): 76, http://doi.org/10.3389/fnbeh.2015.00076.

158 "six basic emotions": Nate Dolensek, Daniel A. Gehrlach, Alexandra S. Klein, and Nadine Gogolla, "Facial Expressions of Emotion States and Their Neuronal Correlates in Mice," *Science* 368, no. 6486 (April 3, 2020): 89–94, https://doi.org/10.1126/science.aaz9468.

158 Planes flying across the Arctic: Graeme Green, "How a Hi-Tech Search for Genghis Khan Is Helping Polar Bears," *Guardian*, April 27, 2021, https://www.theguardian.com/environment/2021/apr/27/polar-bears-genghis-khan-ai-radar-innovations-helping-protect-cubs-aoe.

158 "arguments" between Egyptian fruit bats: Nicola Davis, "Bat Chat: Machine Learning Algorithms Provide Translations for Bat Squeaks," *Guardian*, December 22, 2016, https://www.theguardian.com/science/2016/dec/22/bat-chat-machine-learning-algorithms-provide-translations-for-bat-squeaks.

158 millions of trees: Amy Fleming, "One, Two, Tree: How AI Helped Find Millions of Trees in the Sahara," *Guardian*, January 15, 2021, https://www.theguardian.com/environment/2021/jan/15/how-ai-helped-find-millions-of-trees-in-the-sahara-aoe.

158 predicted volcanic eruptions: Australian Associated Press, "New Zealand Scientists Invent Volcano Warning System," *Guardian*, July 19, 2020, https://www

.theguardian.com/world/2020/jul/20/new-zealand-scientists-invent-volcano-warning-System.

158 fifty-three other species and counting: Wild Me website, https://www.wildme.org/#/.

159 FathomNet is being made publicly available: "FathomNet," Monterey Bay Aquarium Research Institute, https://www.mbari.org/fathomnet/.

159 meta-analysis of one hundred thousand climate change studies: Max Callaghan, Carl-Friedrich Schleussner, Shruti Nath et al., "Machine-Learning-Based Evidence and Attribution Mapping of 100,000 Climate Impact Studies," *Nature Climate Change* 11 (2021): 966–972, https://doi.org/10.1038/s41558-021-01168-6.

159 something called AlphaFold: Andrew W. Senior, Richard Evans, John Jumper et al., "Improved Protein Structure Prediction Using Potentials from Deep Learning," *Nature* 577, no. 7792 (2020): 706–710.

159 "solving intelligence, and then using that to solve everything else": Tom Simonite, "How Google Plans to Solve Artificial Intelligence," *MIT Technology Review*, March 31, 2016, https://www.technologyreview.com/2016/03/31/161234/how-google-plans-to-solve-artificial-intelligence/.

159 a "gargantuan leap": Callaway, "'It Will Change Everything': DeepMind's AI Makes Gigantic Leap in Solving Protein Structures," *Nature* 588, no. 7837 (2020): 203–204.

159 "the core problem has arguably been solved": Ibid.

159 "It's a game changer": Ibid.

160 "force multiplier": Ian Hogarth, conversation with author, May 4, 2020.

161 He found 497 whistles in the recordings: J. Fearey, S. H. Elwen, B. S. James, and T. Gridley, "Identification of Potential Signature Whistles from Free-Ranging Common Dolphins (*Delphinus delphis*) in South Africa," *Animal Cognition* 22, no. 5 (2019): 777–789.

162 she recorded a group of thirteen captive dolphins: Julie N. Oswald, "Bottlenose Dolphin Whistle Repertoires: Size and Stability over Time," presentation at IBAC, University of St. Andrews, UK, September 5, 2019.

162 Julie's AI extracted: Oswald, email communication with the author, November 23, 2021.

165 They bond with their friends: Dudzinski, K., and Ribic, C. "Pectoral fin contact as a mechanism for social bonding among dolphins," February 2017, *Animal Behavior and Cognition,* 4(1):30–48.

165 "a day late and a dollar short": Cheeseman, interview with author, July 29, 2020.

Chapter 10. Machines of Loving Grace

167 "To move forward is to concoct": Edward O. Wilson, *The Diversity of Life* (Cambridge, MA: Belknap Press of Harvard University Press, 1992), 5.

167 was the astronomer Christiaan Huygens: Lane, "The Unseen World: Reflections on Leeuwenhoek."

168 "Scientific discovery requires risk": Nadia Drake, "When Hubble Stared at Nothing for 100 Hours," *National Geographic*, April 24, 2015, https://www.nationalgeographic.com/science/article/when-hubble-stared-at-nothing-for-100-hours.

168 342 pictures: "Discoveries: Hubble's Deep Fields," National Aeronautics and Space Administration, updated October 29, 2021, https://www.nasa.gov/content/discoveries-hubbles-deep-fields.

168 A "cosmic zoo": Hubble explores the origins of modern galaxies, ESA Hubble Media Newsletter, Press Release, August 15, 2013, https://esahubble.org/news/heic1315/.

169 infinite scroll function: Danielle Cohen, "He Created Your Phone's Most Addictive Feature. Now He Wants to Build a Rosetta Stone for Animal Language," *GQ*, July 6, 2021, https://www.gq-magazine.co.uk/culture/article/aza-raskin-interview.

169 "a significant portion of [his] life force": Aza Raskin, email with author, January 3, 2022.

170 the songs of *hundreds* of animals: John P. Ryan, Danelle E. Cline, John E. Joseph et al.,"Humpback Whale Song Occurrence Reflects Ecosystem Variability in Feeding and Migratory Habitat of the Northeast Pacific," *PLoS One* 14, no. 9 (2019): e0222456, https://doi.org/10.1371/journal.pone.0222456.

172 if you fed lots of texts: Tomas Mikolov, Kai Chen, Greg Corrado, and Jeffrey Dean, "Efficient Estimation of Word Representations in Vector Space," arXiv preprint, arXiv:1301.3781 (2013).

172 "You shall know a word by the company it keeps!": John R. Firth, "A Synopsis of Linguistic Theory, 1930–1955," in *Studies in Linguistic Analysis* (Oxford: Blackwell, 1957).

173 here is Britt and Aza's example: "Earth Species Project: Research Direction," GitHub, last modified June 10, 2020, https://github.com/earthspecies/project/blob/master/roadmaps/ai.md.

174 A young researcher named Mikel Artetxe: Mikel Artetxe, Gorka Labaka, Eneko Agirre, and Kyunghyun Cho, "Unsupervised Neural Machine Translation," arXiv:1710.1141 (2017), http://arxiv.org/abs/1710.11041.

174 No examples of translation: Yu-An Chung, Wei-Hung Weng, Schrasing Tong, and James Glass, "Unsupervised Cross-Modal Alignment of Speech and Text Embedding Spaces," arXiv: 1805.07467 (2018), http://arxiv.org/abs/1805.07467.

174 "Imagine being given two entirely": Britt Selvitelle, *Earth Species Project: Research Direction*, Github, June 10, 2020 https://github.com/earthspecies/project/blob/main/roadmaps/ai.md.

177 "most of the smartest people work elsewhere": This quote was originally attributed to Bill Joy in Brent Schlender, "Whose Internet Is It, Anyway?" *Fortune,* December 11, 1995, 120, cited in "The Smartest People in the World Don't All Work for Us. Most of Them Work for Someone Else," Quote Investigator, January 28, 2018, https://quoteinvestigator.com/2018/01/28/smartest/.

178 the "Cocktail Party Problem": Barry Arons, "A Review of the Cocktail Party Effect," *Journal of the American Voice I/O Society* 12, no. 7 (1992): 35–50.

178 their findings were published: Peter C. Bermant, "BioCPPNet: Automatic Bioacoustic Source Separation with Deep Neural Networks," *Scientific Reports* 11 (2021): 23502, https://doi.org/10.1038/s41598-021-02790-2.

179 a hundred miles in a day: Stuart Thornton, "Incredible Journey," *National Geographic*, October 29, 2010, https://www.nationalgeographic.org/article/incredible-journey/.

182 tags revealed these extraordinary underwater acrobatics: David Wiley, Colin Ware, Alessandro Bocconcelli et al., "Underwater Components of Humpback Whale Bubble-Net Feeding Behaviour," *Behaviour* 148, no. 5/6 (2011): 575–602.

182 *bubble-net feeding*: Ibid. http://www.jstor.org/stable/23034261.

183 "a really awesome partnership": Ari Friedlaender, email to author, November 22, 2021.

183 a computer system that a diver can wear: Daniel Kohlsdorf, Scott Gilliland, Peter Presti et al., "An Underwater Wearable Computer for Two Way Human-Dolphin Communication Experimentation," in *Proceedings of the 2013 International Symposium on Wearable Computers* (New York: Association for Computing Machinery, 2013), 147–148, https://doi.org/10.1145/2493988.2494346.

184 Interspecies Internet: "Our Mission," Interspecies Internet, updated April 21, 2021, https://www.interspecies.io/about.

184 "interaction device": Video interview with author, April 11, 2022.

185 capture the glow of sea turtles: Danny Lewis, "Scientists Just Found a Sea Turtle That Glows," *Smithsonian Magazine*, October 1, 2015, https://www.smithsonianmag.com/smart-news/scientists-discover-glowing-sea-turtle-180956789/.

185 gently handle fragile deep-sea animals: Kevin C. Galloway, Kaitlyn P. Becker, Brennan Phillips et al., "Soft Robotic Grippers for Biological Sampling on Deep Reefs," *Soft Robotics* 3, no. 1 (March 17, 2016): 23–33, https://doi.org/10.1089/soro.2015.0019.

185 Their team is huge: Project CETI, https://www.projectceti.org.

185 "to learn how to communicate with a whale": David Gruber, email, December 27, 2021.

186 "from small data to big data": Ibid.

186 Gero's decades of careful listening: Gero, Jonathan Gordon, and Whitehead, "Individualized Social Preferences and Long-Term Social Fidelity Between Social Units of Sperm Whales," *Animal Behaviour* 102 (2015): 15–23, https:/doi.org/10.1016/j.anbehav.2015.01.008.

186 "Core Whale Listening station": "Project Ceti," The Audacious Project Impact 2020, https://impact.audaciousproject.org/projects/project-ceti.

186 "soft robotic fish": Robert K. Katzschmann, Joseph DelPreto, Robert MacCurdy, and Daniela Rus, "Exploration of Underwater Life with an Acoustically Controlled Soft Robotic Fish," *Science Robotics* 3, no. 16 (March 28, 2018): eaar3449, https://doi.org/10.1126/scirobotics.aar3449.

186 The whales will be living in an auditory panopticon: Jacob Andreas, Gašper Beguš, Michael M. Bronstein et al., "Cetacean Translation Initiative: A Roadmap to Deciphering the Communication of Sperm Whales," arXiv preprint, arXiv:2104.08614 (2021).

187 "social network": Ibid.

187 "largest animal behavioural data set" ever gathered: Ibid.

188 "automated machine learning pipelines": Ibid.

188 "the deep wonder of our attempt": "Project Ceti," The Audacious Project.

188 distinguishing between those of different clans and individuals: Gero, Whitehead, and Rendell, "Individual, Unit and Vocal Clan Level Identity Cues in Sperm Whale Codas," *Royal Society Open Science* 3, no. 1 (2016): 150372.

188 the AIs will chart the sperm whale communicative galaxy: Bermant, Bronstein, Robert J. Wood et al., "Deep Machine Learning Techniques for the Detection and Classification of Sperm Whale Bioacoustics," *Scientific Reports* 9 (2019): 12588, https:/doi.org/10.1038/s41598-019-48909-4.

188 "constrain hypothesis space": Andreas, Beguš, Bronstein et al., "Cetacean Translation Initiative."

189 "The important thing to me": Gruber, email, December 27, 2021.

189 The Core Whale Listening Station: Email to author, April 28, 2022.

190 "these AI tools are the invention of the telescope": Raskin, interview, December 17, 2021.

190 "data centric paradigm shift": Andreas, Beguš, Bronstein et al., "Cetacean Translation Initiative."

190 "change our respect for the rest of life entirely": Roger Payne, telephone conversation, December 24, 2021.

191 "Ever since I was a child": Jane Goodall, email to Raskin, August 23, 2020. With permission.

192 "the Animal Internet has the potential": Alexander Pschera, *Animal Internet: Nature and the Digital Revolution*, trans. Elisabeth Lauffer (New York: New Vessel Press, 2016), 11.

192 "What started as a 'hey'": Cheeseman, email communication with author, June 30, 2021.

Chapter 11. Anthropodenial

194 "Animals don't exist in order to teach us things": Helen Macdonald, *Vesper Flights* (New York: Vintage / Penguin Random House, 2021), 255.

194 about forty thousand years ago: Tom Higham, Katerina Douka, Rachel Wood et al., "The Timing and Spatiotemporal Patterning of Neanderthal Disappearance," *Nature* 512, no. 7514 (2014): 306–309.

194 bison and reindeer: Kate Britton, Vaughan Grimes, Laura Niven et al., "Strontium Isotope Evidence for Migration in Late Pleistocene Rangifer: Implications for Neanderthal Hunting Strategies at the Middle Palaeolithic Site of Jonzac, France," *Journal of Human Evolution* 61, no. 2 (2011): 176–185.

195 The stone tips: Marie-Hélène Moncel, Paul Fernandes, Malte Willmes et al., "Rocks, Teeth, and Tools: New Insights into Early Neanderthal Mobility Strategies in South-Eastern France from Lithic Reconstructions and Strontium Isotope Analysis," *PLoS One* 14, no. 4 (2019): e0214925.

195 they made fire: Rosa M. Albert, Francesco Berna, and Paul Goldberg, "Insights on Neanderthal Fire Use at Kebara Cave (Israel) Through High Resolution Study of Prehistoric Combustion Features: Evidence from Phytoliths and Thin Sections," *Quaternary International* 247 (2012): 278–293.

195 they seemed to have religious beliefs: Tim Appenzeller, "Neanderthal Culture: Old Masters," *Nature* 497, no. 7449 (2013): 302.

195 perform major life-saving surgery: Erik Trinkaus and Sébastien Villotte. "External Auditory Exostoses and Hearing Loss in the Shanidar 1 Neandertal," *PLoS One* 12, no. 10 (2017): e0186684.

195 2 percent of our genes: Qiaomei Fu, Mateja Hajdinjak, Oana Teodora Moldovan et al., "An Early Modern Human from Romania with a Recent Neanderthal Ancestor," *Nature* 524, no. 7564 (2015): 216–219.

197 his friend the philosopher Henry More: René Descartes, "To More, 5.ii.1649," in *Selected Correspondence of Descartes*, trans. Jonathan Bennett, Some Texts from Early Modern Philosophy, 2017, https://www.earlymoderntexts.com/assets/pdfs/descartes1619_4.pdf (p. 216).

197 "Cogito, ergo sum": Descartes, *Discourse on the Method of Rightly Conducting One's Reason and of Seeking Truth in the Sciences*, 1637.

198 "merely expressions of their fear": Descartes, "To Cavendish, 23.xi.1646," in *Selected Correspondence of Descartes*, 189.

198 To reason was uniquely human: Colin Allen and Michael Trestman, "Animal Consciousness," *Stanford Encyclopedia of Philosophy Archive*, Winter 2020 edition, ed. Edward N. Zalta, Center for the Study of Language and Information, Stanford University, https://plato.stanford.edu/archives/win2020/entries/consciousness-animal/.

199 cataloguing and organizing nature: Paul S. Agutter and Denys N. Wheatley, *Thinking About Life: The History and Philosophy of Biology and Other Sciences* (Dordrecht, Netherlands: Springer, 2008), 43.

199 Aristotle's *History of Animals*: *Aristotle's History of Animals: In Ten Books*, trans. Richard Cresswell (London: Henry G. Bohn, 1862).

199 twelfth-century botanical texts: Abū ḥanīfah Aḥmad ibn Dāwūd Dīnawarī, *Kitab al-nabat—The Book of Plants*, ed. Bernhard Lewin (Wiesbaden: Franz Steiner, 1974).

199 writings of Saint Albertus: Saint Albertus Magnus, *On Animals: A Medieval Summa Zoologica*, 2 vols., trans. Kenneth M. Kitchell (Baltimore: Johns Hopkins University Press, 1999).

200 "monstrous beast with the hands of a man": Tad Estreicher, "The First Description of a Kangaroo," *Nature* 93, no. 2316 (1914): 60.

201 "an animal that doesn't think it is an animal": Melanie Challenger, *How to Be Animal: A New History of What It Means to Be Human* (Edinburgh: Canongate, 2021).

201 "When I play with my cat": "Apology for Raimond Sebond," chap. 12 in *The Essays of Montaigne, Complete*, trans. Charles Cotton (1887).

201 "the beasts of the field, the fowl of the air and the fish of the sea": Edward L. Thorndike, "The Evolution of the Human Intellect," chap. 7 in *Animal Intelligence* (New York: Macmillan, 1911).

202 "watching and wondering": Nikolaas Tinbergen, "Ethology and Stress Diseases," Nobel Prize in Physiology or Medicine lecture, December 12, 1973, The Nobel Prize, https://www.nobelprize.org/uploads/2018/06/tinbergen-lecture.pdf.

202 dance to their hive-mates: David R. Tarpy, "The Honey Bee Dance Language," NC State Extension, February 23, 2016, https://content.ces.ncsu.edu/honey-bee-dance-language.

203 "I view them more as individuals": Kat Kerlin, "Personality Matters, Even for Squirrels," News and Information, University of Califonia, Davis, September 10, 2021, https://www.ucdavis.edu/curiosity/news/personality-matters-even-squirrels-0.

203 Making tools: Gavin R. Hunt, "Manufacture and Use of Hook-Tools by New Caledonian Crows," *Nature* 379, no. 6562 (1996): 249–251. Robert W. Shumaker, Kristina R. Walkup, and Benjamin B. Beck, *Animal Tool Behavior: The Use and Manufacture of Tools by Animals* (Baltimore: Johns Hopkins University Press, 2011). Vicki Bentley-Condit and E. O. Smith, "Animal Tool Use: Current Definitions and an Updated Comprehensive Catalog," *Behaviour* 147, no. 2 (2010): 185–221.

203 Cooperating to achieve tasks: Tui De Roy, Eduardo R. Espinoza, and Fritz Trillmich, "Cooperation and Opportunism in Galapagos Sea Lion Hunting for Shoaling Fish," *Ecology and Evolution* 11, no. 14 (2021): 9206–9216. Alicia P. Melis, Brian Hare, and Michael Tomasello, "Engineering Cooperation in Chimpanzees: Tolerance Constraints on Cooperation," *Animal Behaviour* 72, no. 2 (2006): 275–286.

203 Planning ahead: Nicola S. Clayton, Timothy J. Bussey, and Anthony Dickinson, "Can Animals Recall the Past and Plan for the Future?" *Nature Reviews Neuroscience* 4, no. 8 (2003): 685–691. William A. Roberts, "Mental Time Travel: Animals Anticipate the Future," *Current Biology* 17, no. 11 (2007): R418–R420.

203 Having menopause: Margaret L. Walker and James G. Herndon, "Menopause in Nonhuman Primates?" *Biology of Reproduction* 79, no. 3 (2008): 398–406. Rufus A. Johnstone and Michael A. Cant, "The Evolution of Menopause in Cetaceans and Humans: The Role of Demography," *Proceedings of the Royal Society B: Biological Sciences* 277, no. 1701 (2010): 3765–3771, https://doi.org/10.1098/rspb.2010.0988.

203 Understanding abstract concepts: Jennifer Vonk, "Matching Based on Biological Categories in Orangutans (*Pongo abelii*) and a Gorilla (*Gorilla gorilla gorilla*)," *PeerJ* 1 (2013): e158. Pepperberg, "Abstract Concepts: Data from a Grey Parrot," *Behavioural Processes* 93 (2013): 82–90, https://doi.org/10.1016/j.beproc.2012.09.016. Herman, Adam A. Pack, and Amy M. Wood, "Bottlenose Dolphins Can Generalize Rules and Develop Abstract Concepts," *Marine Mammal Science* 10, no. 1 (1994): 70–80, https://doi.org/10.1111/j.1748-7692.1994.tb00390.x.

203 Memorizing hundreds of words: John W. Pilley and Alliston K. Reid, "Border Collie Comprehends Object Names as Verbal Referents," *Behavioural Processes* 86, no. 2 (2011): 184–195, https://doi.org/10.1016/j.beproc.2010.11.007. Pepperberg, "Cognitive and Communicative Abilities of Grey Parrots," *Current Directions* *in Psychological Science* 11, no. 3 (2002): 83–87. R. Allen Gardner and Beatrice T. Gardner, "Teaching Sign Language to a Chimpanzee," *Science* 165, no. 3894 (August 15, 1969): 664–672. Francine G. Patterson, "The Gestures of a Gorilla: Language Acquisition in Another Pongid," *Brain and Language* 5, no. 1 (1978): 72–97.

203 Remembering long number sequences: Nobuyuki Kawai and Tetsuro Matsuzawa, "Numerical Memory Span in a Chimpanzee," *Nature* 403, no. 6765 (2000): 39–40.

203 Doing simple mathematics: Pepperberg, "Grey Parrot Numerical Competence: A Review," *Animal Cognition* 9, no. 4 (2006): 377–391. Sara Inoue and Matsuzawa, "Working Memory of Numerals in Chimpanzees," *Current Biology* 17, no. 23 (2007): R1004–R1005.

203 Recognizing human faces: Cait Newport, Guy Wallis, Yarema Reshitnyk, and Ulrike E. Siebeck, "Discrimination of Human Faces by Archerfish (*Toxotes chatareus*)," *Scientific Reports* 6, no. 1 (2016): 1–7. Franziska Knolle, Rita P. Goncalves, and A. Jennifer Morton, "Sheep Recognize Familiar and Unfamiliar Human Faces from Two-Dimensional Images," *Royal Society Open Science* 4, no. 11 (2017): 171228. Anaïs Racca, Eleonora Amadei, Séverine Ligout et al., "Discrimination of Human and Dog Faces and Inversion Responses in Domestic Dogs (*Canis familiaris*)," *Animal Cognition* 13, no. 3 (2010): 525–533.

203 Making and having friends: Jorg J. M. Massen and Sonja E. Koski, "Chimps of a Feather Sit Together: Chimpanzee Friendships Are Based on Homophily in Personality," *Evolution and Human Behavior* 35, no. 1 (2014): 1–8. Robin Dunbar, "Do Animals Have Friends, Too?" *New Scientist*, May 21, 2014, https://www.newscientist.com/article/mg22229700-400-friendship-do-animals-have-friends-too/. Michael N. Weiss, Daniel Wayne Franks, Deborah A. Giles et al., "Age and Sex Influence Social Interactions, but Not Associations, Within a Killer Whale Pod," *Proceedings of the Royal Society B: Biological Sciences* 288, no. 1953 (2021): 1–28.

203 Kissing with tongues: Joseph H. Manson, Susan Perry, and Amy R. Parish, "Nonconceptive Sexual Behavior in Bonobos and Capuchins," *International Journal of Primatology* 18, no. 5 (1997): 767–786. Benjamin Lecorps, Daniel M. Weary, and Marina A. G. von Keyserlingk, "Captivity-Induced Depression in Animals," *Trends in Cognitive Sciences* 25, no. 7 (2021): 539–541.

203 Experiencing mental illnesses: Jaime Figueroa, David Solà-Oriol, Xavier Manteca et al., "Anhedonia in Pigs? Effects of Social Stress and Restraint Stress on Sucrose Preference," *Physiology & Behavior* 151 (2015): 509–515.

203 Grieving: Teja Brooks Pribac, "Animal Grief," *Animal Studies Journal* 2, no. 2 (2013): 67–90. Carl Safina, "The Depths of Animal Grief," *Nova*, PBS, July 8, 2015, https://www.pbs.org/wgbh/nova/article/animal-grief/.

203 Using syntax: Zuberbühler, "Syntax and Compositionality in Animal Communication," *Philosophical Transactions of the Royal Society B: Biological Sciences* 375, no. 1789 (2020): 20190062. Suzuki, David Wheatcroft, and Michael Griesser, "The Syntax–Semantics Interface in Animal Vocal Communication," *Philosophical Transactions of the Royal Society B: Biological Sciences* 375, no. 1789 (2020): 20180405. Robert C. Berwick, Kazuo Okanoya, Gabriel J. L. Beckers, and Johan J. Bolhuis, "Songs to Syntax: The Linguistics of Birdsong," *Trends in Cognitive Sciences* 15, no. 3 (2011):113–121, https://doi.org/10.1016/j.tics.2011.01.002, PMID: 21296608.

203 Falling in "love": Marc Bekoff, "Animal Emotions: Exploring Passionate Natures: Current Interdisciplinary Research Provides Compelling Evidence That Many Animals Experience Such Emotions as Joy, Fear, Love, Despair, and Grief—We Are Not Alone," *BioScience* 50, no. 10 (2000): 861–870. Pepperberg, "Functional Vocalizations by an African Grey Parrot (*Psittacus erithacus*)," *Zeitschrift für Tierpsychologie* 55, no. 2 (1981): 139–160.

203 Feeling jealous: Amalia P. M. Bastos, Patrick D. Neilands, Rebecca S. Hassall et al., "Dogs Mentally Represent Jealousy-Inducing Social Interactions," *Psychological Science* 32, no. 5 (2021): 646–654.

203 Accurately mimicking human speech: Pepperberg, "Vocal Learning in Grey Parrots: A Brief Review of Perception, Production, and Cross-Species Comparisons," *Brain and Language* 115, no. 1 (2010): 81–91. Abramson, Hernández-Lloreda, Lino García et al., "Imitation of Novel Conspecific and Human Speech Sounds in the Killer Whale (*Orcinus orca*)," *Proceedings of the Royal Society B: Biological Sciences* 285, no. 1871 (2018): 20172171, https://doi.org/10.1098/rspb.2017.2171; erratum in *Proceedings of the Royal Society B: Biological Sciences* 285, no. 1873 (2018): 20180297, https://doi.org/10.1098/rspb.2018.0287. Angela S. Stoeger et al., "An Asian Elephant Imitates Human Speech," *Current Biology* 22, no. 22 (2012): P2144–P2148, https://doi.org/10.1016/j.cub.2012.09.022.

203 Laughing: Marina Davila-Ross, Michael J. Owren, and Elke Zimmermann, "Reconstructing the Evolution of Laughter in Great Apes and Humans," *Current Biology* 19, no. 13 (2009): 1106–1111. Davila-Ross, Goncalo Jesus, Jade Osborne, and Kim A. Bard, "Chimpanzees (*Pan troglodytes*) Produce the Same Types of 'Laugh Faces' When They Emit Laughter and When They Are Silent," *PLoS One* 10, no. 6 (2015): e0127337.

203 Experiencing awe, wonder, or even "spiritual" experiences: Kevin Nelson, *The Spiritual Doorway in the Brain: A Neurologist's Search for the God Experience* (New York: Dutton/Penguin, 2011). Barbara J. King, "Seeing Spirituality in Chimpanzees," *Atlantic*, March 29, 2016, https://www.theatlantic.com/science/archive/2016/03/chimpanzee-spirituality/475731/.

203 Feeling pain: T. C. Danbury, C. A. Weeks, A. E. Waterman-Pearson et al., "Self-Selection of the Analgesic Drug Carprofen by Lame Broiler Chickens," *Veterinary Record* 146, no. 11 (2000): 307–311. Earl Carstens and Gary P. Moberg, "Recognizing Pain and Distress in Laboratory Animals," *ILAR Journal* 41, no. 2 (2000): 62–71. Liz Langley, "The Surprisingly Humanlike Ways Animals Feel Pain," *National Geographic*, December 3, 2016, https://www.nationalgeographic.com/animals/article/animals-science-medical-pain.

204 Feeling pleasure: Michel Cabanac, "Emotion and Phylogeny," *Journal of Consciousness Studies* 6, no. 6–7 (1999): 176–190. Jonathan Balcombe, "Animal

Pleasure and Its Moral Significance," *Applied Animal Behaviour Science* 118, no. 3–4 (2009): 208–216.

204 Gossiping: Ipek G. Kulahci, Daniel I. Rubenstein, and Ghazanfar, "Lemurs Groom-at-a-Distance Through Vocal Networks," *Animal Behaviour* 110 (2015): 179–186. Kieran C. R. Fox, Michael Muthukrishna, and Susanne Shultz, "The Social and Cultural Roots of Whale and Dolphin Brains," *Nature Ecology & Evolution* 1, no. 11 (2017): 1699–1705.

204 Killing for "pleasure": Kimberley Hickock, "Rare Footage Shows Beautiful Orcas Toying with Helpless Sea Turtles," *Live Science*, September 20, 2018, https://www.livescience.com/63622-orca-spins-sea-turtle.html.

204 Playing: Fox et al., "The Social and Cultural Roots of Whale and Dolphin Brains." Gordon M. Burghardt, *The Genesis of Animal Play: Testing the Limits* (Cambridge, MA: MIT Press, 2006).

204 Exhibiting morality: De Waal, *The Age of Empathy: Nature's Lessons for a Kinder Society* (London: Souvenir Press, 2010). Susana Monsó, Judith Benz-Schwarzburg, and Annika Bremhorst, "Animal Morality: What It Means and Why It Matters," *Journal of Ethics* 22, no. 3 (2018), 283–310, https://doi.org/10.1007/s10892-018-9275-3.

204 Demonstrating a sense of fairness: Sarah F. Brosnan and de Waal, "Evolution of Responses to (Un)fairness," *Science* 346, no. 6207 (September 18, 2014), https://doi.org/10.1126/science.1251776. Claudia Wascher, "Animals Know When They Are Being Treated Unfairly (and They Don't Like It)," The Conversation, Phys.org, February 22, 2017, https://phys.org/news/2017-02-animals-unfairly-dont.html.

204 Performing altruistic behavior: Indrikis Krams, Tatjana Krama, Kristine Igaune, and Raivo Mänd, "Experimental Evidence of Reciprocal Altruism in the Pied Flycatcher," *Behavioral Ecology and Sociobiology* 62, no. 4 (2008): 599–605. De Waal, "Putting the Altruism Back into Altruism: The Evolution of Empathy," *Annual Review of Psychology* 59 (2008): 279–300.

204 Making art: Lesley J. Rogers and Gisela Kaplan, "Elephants That Paint, Birds That Make Music: Do Animals Have an Aesthetic Sense?" *Cerebrum 2006: Emerging Ideas in Brain Science* (2006): 1–14. Jason G. Goldman, "Creativity: The Weird and Wonderful Art of Animals," BBC, July 23, 2014, https://www.bbc.com/future/article/20140723-are-we-the-only-creative-species.

204 Keeping time: Ferris Jabr, "The Beasts That Keep the Beat," *Quanta Magazine*, March 22, 2016, https://www.quantamagazine.org/the-beasts-that-keep-the-beat-20160322/.

204 dancing: Russell A. Ligon, Christopher D. Diaz, Janelle L. Morano et al., "Evolution of Correlated Complexity in the Radically Different Courtship Signals of Birds-of-Paradise," *PLoS Biology* 16, no. 11 (2018): e2006962. Emily Osterloff, "Best Foot Forward: Eight Animals That Dance to Impress," Natural History Museum (London), March 12, 2020, https://www.nhm.ac.uk/discover/animals-that-dance-to-impress.html.

204 Laughing, including when tickled: Jaak Panksepp and Jeffrey Burgdorf, "50-kHz Chirping (Laughter?) in Response to Conditioned and Unconditioned Tickle-Induced Reward in Rats: Effects of Social Housing and Genetic Variables," *Behavioural Brain Research* 115, no. 1 (2000): 25–38.

204 Weighing up probabilities: James A. R. Marshall, Gavin Brown, and Andrew N. Radford, "Individual Confidence-Weighting and Group Decision-Making," *Trends in Ecology & Evolution* 32, no. 9 (2017): 636–645. Davis and Eleanor Ainge Roy, "Study Finds Parrots Weigh Up Probabilities to Make Decisions," *Guardian*, March 3, 2020, https://www.theguardian.com/science/2020/mar/03/study-finds-parrots-weigh-up-probabilities-to-make-decisions.

204 Emotional contagion: Ana Pérez-Manrique and Antoni Gomila, "Emotional Contagion in Nonhuman Animals: A Review," *Wiley Interdisciplinary Reviews: Cognitive Science* 13, no. 1 (2022): e1560. Julen Hernandez-Lallement, Paula Gómez-Sotres, and Maria Carrillo, "Towards a Unified Theory of Emotional Contagion in Rodents—A Meta-analysis," *Neuroscience & Biobehavioral Reviews* (2020).

204 Rescuing and comforting one another: A. Roulin, B. Des Monstiers, E. Ifrid et al., "Reciprocal Preening and Food Sharing in Colour-Polymorphic Nestling Barn Owls," *Journal of Evolutionary Biology* 29, no. 2 (2016): 380–394. Pitman, Volker B. Deecke, Christine M. Gabriele et al., "Humpback Whales Interfering When Mammal-Eating Killer Whales Attack Other Species: Mobbing Behavior and Interspecific Altruism?" *Marine Mammal Science* 33, no. 1 (2017): 7–58.

204 Displaying accents: Philip Hunter, "Birds of a Feather Speak Together: Understanding the Different Dialects of Animals Can Help to Decipher Their Communication," *EMBO Reports* 22, no. 9 (2021): e53682. Antunes, Tyler Schulz, Gero et al., "Individually Distinctive Acoustic Features in Sperm Whale Codas," *Animal Behaviour* 81, no. 4 (2011): 723–730, https://doi.org/10.1016/j.anbehav.2010.12.019.

204 Having and transmitting cultures: Bennett G. Galef, "The Question of Animal Culture," *Human Nature* 3, no. 2 (1992): 157–178. Andrew Whiten, Goodall, William C. McGrew et al., "Cultures in Chimpanzees," *Nature* 399, no. 6737 (1999): 682–685. Michael Krützen, Erik P. Willems, and Carel P. van Schaik, "Culture and Geographic Variation in Orangutan Behavior," *Current Biology* 21, no. 21 (2011): 1808–1812. Whitehead and Rendell, *The Cultural Lives of Whales and Dolphins* (Chicago: University of Chicago Press, 2015).

204 Predicting the intentions of others: Fumihiro Kano, Christopher Krupenye, Satoshi Hirata et al., "Great Apes Use Self-Experience to Anticipate an Agent's Action in a False-Belief Test," *Proceedings of the National Academy of Sciences of the United States of America* 116, no. 42 (2019): 20904–20909.

204 Intentionally intoxicating themselves: Jorge Juarez, Carlos Guzman-Flores, Frank R. Ervin, and Roberta M. Palmour, "Voluntary Alcohol Consumption in Vervet Monkeys: Individual, Sex, and Age Differences," *Pharmacology,*

Biochemistry, and Behavior 46, no. 4 (1993): 985–988. Christie Wilcox, "Do Stoned Dolphins Give 'Puff Puff Pass' A Whole New Meaning?" *Discover*, December 30, 2013, https://www.discovermagazine.com/planet-earth/do-stoned-dolphins-give-puff-puff-pass-a-whole-new-meaning#.VIHlOWTF_OZ.

204 Manipulating and deceiving others: Hare, Call, and Tomasello, "Chimpanzees Deceive a Human Competitor by Hiding," *Cognition* 101, no. 3 (2006): 495–514. Kazuo Fujita, Hika Kuroshima, and Saori Asai, "How Do Tufted Capuchin Monkeys (*Cebus apella*) Understand Causality Involved in Tool Use?" *Journal of Experimental Psychology: Animal Behavior Processes* 29, no. 3 (2003): 233.

204 "anthropodenial": De Waal, "Are We in Anthropodenial?" *Discover*, July 1997.

205 Elephants will turn: Karen McComb, Lucy Baker, and Cynthia Moss, "African Elephants Show High Levels of Interest in the Skulls and Ivory of Their Own Species," *Biology Letters* 2, no. 1 (2006): 26–28.

205 like the orca mother: Bopha Phorn, "Researchers Found Orca Whale Still Holding On to Her Dead Calf 9 Days Later," ABC News, August 1, 2018, https://abcnews.go.com/US/researchers-found-orca-whale-holding-dead-calf-days/story?id=56965753.

206 "no reason to think grief would be restricted": Colin Allen and Trestman, "Animal Consciousness."

206 toy with prey species: Hickock, "Rare Footage Shows Beautiful Orcas Toying with Helpless Sea Turtles."

206 ride in the bow waves: Bernd Würsig, "Bow-Riding," in *Encyclopedia of Marine Mammals*, 2nd ed., ed. William F. Perrin, Würsig, and J. G. M. Thewissen (London: Academic Press, 2009).

206 harbor porpoises to death: Peter Fimrite, "'Porpicide': Bottlenose Dolphins Killing Porpoises," *SFGate*, September 17, 2011, https://www.sfgate.com/news/article/Porpicide-Bottlenose-dolphins-killing-porpoises-2309298.php.

206 care for their sick and disabled: Justine Sullivan, "Disabled Killer Whale Survives with Help from Its Pod," Oceana, May 21, 2013, https://usa.oceana.org/blog/disabled-killer-whale-survives-help-its-pod/.

206 those strange orcas: Aimee Gabay, "Why Are Orcas 'Attacking' Fishing Boats off the Coast of Gibraltar?" *New Scientist*, September 15, 2021, https://www.newscientist.com/article/mg25133521-100-why-are-orcas-attacking-fishing-boats-off-the-coast-of-gibraltar/.

207 "talking in their sleep": Sara Reardon, "Do Dolphins Speak Whale in Their Sleep?" *Science*, January 20, 2012, https://www.science.org/content/article/do-dolphins-speak-whale-their-sleep.

207 "a complicated case": Oliver Milman, "Anthropomorphism: How Much Humans and Animals Share Is Still Contested," *Guardian*, January 15, 2016, https://www.theguardian.com/science/2016/jan/15/anthropomorphism-danger-humans-animals-science.

207 "It's exactly like white supremacy": Author's interview with Roger Payne, New York City, April 7, 2019.

208 "The future is already here": "The Future Has Arrived—It's Just Not Evenly Distributed Yet," Quote Investigator, https://quoteinvestigator.com/2012/01/24/future-has-arrived/.

208 in 2012, a convention of scientists: Bekoff, "Scientists Conclude Nonhuman Animals Are Conscious Beings," *Psychology Today*, August 10, 2012, https://www.psychologytoday.com/gb/blog/animal-emotions/201208/scientists-conclude-nonhuman-animals-are-conscious-beings.

208 "that humans are not unique in possessing the neurological substrates": "The Cambridge Declaration on Consciousness," Francis Crick Memorial Conference, July 7, 2012, http://fcmconference.org/img/CambridgeDeclarationOnConsciousness.pdf.

208 "examples of higher levels of consciousness": Pierre Le Neindre, Emilie Bernard, Alain Boissy et al., "Animal Consciousness," *EFSA Supporting Publications* 14, no. 4 (2017): 1196E.

208 the most viral political news story: Jim Waterson, "How a Misleading Story About Animal Sentience Became the Most Viral Politics Article of 2017 and Left Downing Street Scrambling," *BuzzFeed News*, November 25, 2017, https://www.buzzfeed.com/jimwaterson/independent-animal-sentience.

208 It was shared half a million times: Yas Necati, "The Tories Have Voted That Animals Can't Feel Pain as Part of the EU Bill, Marking the Beginning of our Anti-science Brexit," *Independent*, November 20, 2017, https://www.independent.co.uk/voices/brexit-government-vote-animal-sentience-can-t-feel-pain-eu-withdrawal-bill-anti-science-tory-mps-a8065161.html.

209 Animal Sentience Bill: Animal Welfare (Sentience) Bill [HL], Government Bill, Originated in the House of Lords, Session 2021–22, UK Parliament (website), https://bills.parliament.uk/bills/2867.

209 "Animals officially have feelings": Good Morning Britain (@GMB), "Animals officially have feelings. Is it time to stop eating them?" Twitter, May 13, 2021, https://twitter.com/GMB/status/1392744824705536002?s=20&t=YOjJxUydjkTXFYXIn2bu4A.

209 This "speaks to our relationship with all the life around us": Motion No. 2018-268, In the Matter of Nonhuman Rights Project, Inc., on Behalf of Tommy, Appellant, v. Patrick C. Lavery, & c., et al., Respondents and In the Matter of Nonhuman Rights Project, Inc., on Behalf of Kiko, Appellant, v. Carmen Presti et al., Respondents, State of New York Court of Appeals, decided May 8, 2018, https://www.nycourts.gov/ctapps/Decisions/2018/May18/M2018-268opn18-Decision.pdf.

209 desire for freedom, he told me: Author's interview with Steven Wise, April 28, 2019 .

210 "Because we are part of nature": Greta Thunberg (@GretaThunberg), "Our relationship with nature is broken. But relationships can change. When we protect nature—we are nature protecting itself," Twitter, May 22, 2021, https://twitter.com/GretaThunberg/status/1396058911325790208?s=20&t=Zm6rbSY1ZMfypUCmpNO9DA.

210 they consume far more krill: Matthew S. Savoca et al., "Baleen Whale Prey Consumption Based on High-Resolution Foraging Measurements," *Nature* 599 (2021): 85–90, https://www.nature.com/articles/s41586-021-03991-5.

210 the lifetime value of the average baleen whale: Ralph Chami, Thomas Cosimano, Connel Fullenkamp, and Sena Oztosun, "Nature's Solution to Climate Change," *Finance & Development* 56, no. 4 (December 2019), https://www.imf.org/external/pubs/ft/fandd/2019/12/pdf/natures-solution-to-climate-change-chami.pdf.

210 80 percent of wild marine mammals have been lost: Carrington, "Humans Just 0.01% of All Life."

211 "O wad some Pow'r the giftie gie us": Robert Burns, "To a Louse," 1786, Complete Works, Burns Country, http://www.robertburns.org/works/97.shtml.

Chapter 12. Dances with Whales

212 "Not known, because not looked for": Thomas Stearns Eliot, "Little Gidding," in *Four Quartets* (New York: Harcourt, Brace, 1943).

215 has witnessed in its entirety: Nicola Ransome, Lars Bejder, Micheline Jenner et al., "Observations of Parturition in Humpback Whales (*Megaptera novaeangliae*) and Occurrence of Escorting and Competitive Behavior Around Birthing Females," *Marine Mammal Science*, epub September 7, 2021, https://doi.org/10.1111/mms.12864.

217 9.8 megajoules of energy: Segre et al., "Energetic and Physical Limitations on the Breaching Performance of Large Whales."

217 enough to power a home for a day: "Average Gas & Electricity Usage in the UK—2020," Smarter Business, https://smarterbusiness.co.uk/blogs/average-gas-electricity-usage-uk/.

220 by what they called "the sublime": Robert Macfarlane, *Mountains of the Mind: A History of a Fascination* (London: Granta Books, 2009), 75.

221 "True, you learn yourself to be a blip": Ibid.

223 but printers can: Shubham Agrawal, "How Does a Printer Work?—Part I," Medium, March 18, 2020, https://medium.com/@sa159871/how-does-a-printer-work-de0404e3b388.

223 the code that makes a person: "Base Pair," National Human Genome Research Institute, https://www.genome.gov/genetics-glossary/Base-Pair.

223 to form your lover cloned anew: Francisco J. Ayala, "Cloning Humans? Biological, Ethical, and Social Considerations," *Proceedings of the National Academy of Sciences of the United States of America* 112, no. 29 (2015): 8879–8886.

223 called the Human Genome Project: Judith L.Fridovich-Keil, "Human Genome Project," *Encyclopaedia Britannica*, February 27, 2020, https://www.britannica.com/event/Human-Genome-Project.

224 $1,000 to map an entire human genome: "DNA Sequencing Fact Sheet," National Human Genome Research Institute, https://www.genome.gov/about-genomics/fact-sheets/DNA-Sequencing-Fact-Sheet.

224 our DNA with our close cousins: The Chimpanzee Sequencing and Analysis Consortium (Tarjei Mikkelsen, LaDeana Hillier, Evan Eichler et al.), "Initial Sequence of the Chimpanzee Genome and Comparison with the Human Genome," *Nature* 437, no. 7055 (2005): 69–87.

226 here's what they said: Earth Species, email to author, October 16, 2020.

226 "the past is a regret, the future an experiment": Stephen Brennan, ed., *Mark Twain on Common Sense: Timeless Advice and Words of Wisdom from America's Most-Revered Humorist* (New York: Skyhorse Publishing, 2014), 6.

227 vast revenues of the pet industry: "Pet Care Market Size, Share and COVID-19 Impact Analysis, by Product Type (Pet Food Products, Veterinary Care, and Others), Pet Type (Dog, Cat, and Others), Distribution Channel (Online and Offline), and Regional Forecast, 2021–2028," Fortune Business Insights, February 2021, https://www.fortunebusinessinsights.com/pet-care-market-104749. ("The global pet care market size was USD 207.90 billion in 2020.")

227 the arms industry: "Financial Value of the Global Arms Trade," Stockholm International Peace Research Institute, https://www.sipri.org/databases/financial-value-global-arms-trade. ("For example, the estimate of the financial value of the global arms trade for 2019 was at least $118 billion.")

PHOTO CREDITS

Front Matter

1. Sarah A. King

Introduction

2. Jan Verkolje (via Delft University of Technology)
3. Antonie van Leeuwenhoek

Chapter 1. Enter, Pursued by Whale

4. Michael Sack
5. Larry Plants / Storyful
6. Michael Sack
7. Ru Mahoney

Chapter 2. A Song in the Ocean

8. Roger Payne / Ocean Alliance
9. *Science*
10. National Astronomy and Ionosphere Center
11. J. Gregory Sherman

Chapter 3. The Law of the Tongue

12. Claude Rives / Eric Parmentier
13. Public domain
14. Mark D. Scherz
15. Jodi Frediani
16. Public domain
17. Eden Killer Whale Museum

Chapter 4. The Joy of Whales

18. Tom Mustill
19. Sinclair Broadcast Group
20. Anna Ashcroft / Windfall Films
21. Anna Ashcroft / Windfall Films
22. Anna Ashcroft / Windfall Films

Chapter 5. "Some Sort of Stupid, Big Fish"

23. Heidi Whitehead / Texas Marine Mammal Stranding Network
24. Tom Mustill
25. FalseKnees
26. Boris Dimitrov / Public Domain
27. Jillian Morris

Chapter 6. The Search for Animal Language

28. Hernan Segui
29. Andrew Davidhazy
30. Tecumseh Fitch
31. Liz Rubert-Pugh
32. William Munoz
33. The Gorilla Foundation
34. Elaine Miller Bond

Chapter 7. Deep Minds: Cetacean Culture Club

35. Augusto Leandro Stanzani / ardea.com
36. Lilly Estate
37. Walt Disney World Corporation
38. Diana Reiss
39. Eric A. Ramos and Diana Reiss

Chapter 8. The Sea Has Ears

40. Adam Ernster
41. Patrick Hart / LOHE
42. Tom Mustill
43. Ann Tanimoto-Johnson
44. Tom Mustill

Chapter 9. Animalgorithms

45. James T .Tanner and Tensas River National Wildlife Refuge, U.S. Fish and Wildlife Service, Ivory-Billed Woodpecker Records (Mss. 4171), Louisiana and Lower Mississippi Valley Collections, Louisiana State University Libraries, Baton Rouge, Louisiana, USA.
46. Jörg Rychen
47. Public Domain
48. Kate Spencer / Happywhale
49. Kate Cummings / Happywhale
50. Julia Kuhl

51. X, the moonshot factory
52. Vincent Janik, University of St. Andrews

Chapter 10. Machines of Loving Grace

53. R. Williams (STScI), the Hubble Deep Field Team and NASA/ESA
54. Earth Species Project
55. Earth Species Project
56. Ari Friedlaender
57. Ari Friedlaender
58. David Wiley / Colin Ware / Ari Friedlaender
59. David Gruber
60. Alex Boersma
61. Joseph DelPreto / MIT CSAIL
62. Ted Cheeseman
63. Aleksander Nordahl

Chapter 11. Anthropodenial

64. Pedro Saura
65. Public Domain
66. Robin W. Baird / Cascadia Research

Chapter 12. Dances with Whales

67. Anuar Patjana Floriuk
68. Tom Mustill
69. Gene Flipse
70. Jeff Pantukhoff
71. Luke Moss

INDEX

NOTE: Page references in *italics* indicate photographs

Aboriginal people, 44–49, 51–52
Acacia trees, 38–39
Adaptation, 82–83
Ahla (baboon), 42
AI. *See* Artificial intelligence
Akeakamai (dolphin), 117–18
Alarm calls, 104–5, 174–75
  of prairie dogs, 105–7, *106*
Alaska, 4, 124, 143, 176–77
Albertus, Saint, 199
Alex (parrot), *100,* 100–102, 116, 117
Algae, 38
Aliperti, Jaclyn, 203
Allen, Arthur, 145
Alliances, 38
Allman, John, 81
AlphaFold, 159
AlQuriashi, Mohammed, 159
Altruism, 165, 204
Amazon Web Services, 185
Ambergris, 59–60
American Sign Language (ASL), 89, 98–99
Animal cognition, 76, 77–79, 81–82, 85, 101, 103, 117, 122
Animal communication, 87–109, 125. *See also specific animals*
  alarm calls. *See* Alarm calls
  captivity experiments, 97–103, 116–17
  decoding. *See* Decoding animal communication
  Fitch's research , 94–96
  genetics of, 224–25
  in prairie dogs, 105–7, *106*
  vocal learning, 96–103, 113–14
Animal consciousness, 78, 80, 115, 120, 122, 208, 222–23
Animalcules, xiv–xv, *xv,* 167, 168–69, 225
Animal intelligence, 78–79, 83–84
  brain size and, 76–77
Animal language, 89–94
  use of term, 89–91
Animal rights, 208–9
Animal Sentience Bill of 2022, 209
Antarctica, 17, 30, 59, 183, 210
Anthropocene, 29
Anthropocentrism, 79, 196–204
Anthropodenialism, 204–6, 207
Anthropomorphism, 14–15, 83, 206–7
Ants, 38, 39, 76
Apes. *See* Primates
Apollo program, 225
Apple Macintosh, 169
Apps, 157, 191–92, 227
Aristotle, 198–99
Arrival (movie), 113
Artetxe, Mikel, 174
Artificial intelligence (AI), xviii, 154–66, 227
  apps, 157, 191–92, 227
  Cetacean Translation Initiative, 184–91, *187*
  Earth Species Project, 176–78, 183, 191

Artificial intelligence (AI) (*cont.*)
Oswald's dolphin research, 161–63, 171, 222
Raskin and Selvitelle's research, 169–74, 176–78, 183, 190, 226
Astronomy, 199
Attenborough, David, 29
"Attention extraction economy," 169
AudioMoths, 135, 137, 138, 139
Australia, 70, 107
Aboriginal people, 44–49, 51–52
Automata, 198
Azores, 68–69

Baboons, 40–42, *41,* 48, 102
Baleen, 30
Baleen whales, 43–45, 47, 210
Barnacles, whale, 6–7, 9, 12, 37, 53, 153–54
Bats, 91, 97, 147, 158, 177, 178
Bayley (dolphin), 121
BBC Natural History Unit, 15–16
Beach strandings, 50–51, 60–61, 69
Beale, Thomas, 65
Beavers, 84
Beehives, 20, 37
Behavioral biology. *See* Ethology
Behavioral ecology, 202–3
Behaviors
play, 12, 111–12, 114, 201
pointing, 112–13
self-directed, 119–21
Beluga whales, 15, 68, 97–98, 177
Big data, xvii–xviii, 17–18, 143–44, 151–52, 184–90
Cetacean Translation Initiative, 184–91, *187*
FathomNet, 158–59
Happywhale, 153–58, 162, 165, 171–72, 192, *192*
use in author's breaching incident, 18, 152–58, *156*
Big Lebowski, The (movie), 179
Biology, xviii, 199, 200–203
Biomass, 29
Biophony, 149–50
Birds
AI-powered apps for identifying, 191
brain size, 76, 78
extinctions, 131
Birds of paradise, 91, 147
Bird vocalization (birdsong), 26, 38–39, 104, 133–38, 152
AI-powered apps for identifying, 191
Engesser and building blocks, 107–8
recordings, 133–38, 140–41, 145, 152
sex differences, 150–51
vocal learning, 96–97
Birth and birthing, 215, 219–20
Birth scars, 215
Blossom (app), 191
Blowhole, 63, 219
Blubber, 55, 57
Blue whales, 25–26, 30–31, 43, 87, 111–12
Body hair, 55
Bonds (bonding), 69, *202*
Bonobos, 94, 98–99, 121
Bottlenose dolphins, 14, 63, 120–21, 122–23, 175, 206
brains of, 82, *84*
human symbioses, 49–51
Bowhead whales, 26
Brain, 71–86
communication and, 89, 96–97
Neanderthal skulls, 195
regions of, 76–79, 80–82
Reidenberg's scans, 71–75, *75,* 77–80, 82–83, 86
size of, 76–77, 84, *84*
Brautigan, Richard, 128
Breaching, 5, 7, 12–13, 217
Breaching incident of author, 5–13, 29–30, 31, 152–53
news cycle reporting on, 11, 13–14
Plant's photo, *10,* 10–11, *11,* 153

Prime Suspect, 18, 53, 55, 157, 170
use of big data, 18, 152–58, *156*
Brexit, 208–9
Bruno, Giordano, 199
Bryde, Johan, 59
Bryde's whale, 59
"Bubble nets," 182
Bubble rings, *112,* 112–13
Buddha, 20
Budgies, 150
Budongo Forest, 99–100
Burns, Robert, 211
"Burst pulses," 162

Calgacus, 29
Calls. *See* Alarm calls; Vocalizations
Cambrian explosion, 178
Cambridge Declaration of Consciousness, 208
Cameron, David, 13
Capitol Records, 87
Captive animal experiments, 100–103, 162
Reiss and dolphins, 112–15, 116–17, 119, *120,* 120–21, 123, 125–27
Carson, Johnny, 27
Carson, Rachel, xiii
Cats, 121, 149, 150
Cattle egrets, 37
Cave art, 195, *197*
Center for Humane Technology, 169
Central Washington University, 99
Cerebral cortex, 76–77
Cerf, Vint, 184
Cetaceans. *See also* Dolphins; Porpoises; Whales
classification of, 42–45, 160–61
death and grief, 205–6
etymology, 42
evolution of, 6, 43–45, 55
human symbioses. *See* Human-cetacean symbioses
senses of, 38, 61–62
strandings, 50–51, 60–61, 69
Cetacean brains. *See* Whale brain
Cetacean communication. *See* Dolphin communication; Whale communication
"Cetaceans Have Complex Brains for Complex Cognition" (Marino), 82–83
Cetacean Strandings Investigation Programme (CSIP), 56
Cetacean Translation Initiative (CETI), 184–91, *187,* 222
Challenger, Melanie, 201
Chasing Doctor Dolittle (Slobodchikoff), 103–4, 106
Cheeseman, Ted, 153–58, 165, 192, *192,* 222
Chestnut-crowned babblers, 107
Chickens, 29, 105
Chimpanzees, 203
brain size, 76, 77, 78, 80
communication, 94–95, 96, 98–100
genome, 224–25
legal rights of, 209
Mirror Self Recognition test, 121
Christianity, 198–200
Cicadas, 147
Cingulate cortex, 81
Clarke, Arthur C., 30–31
Clarke, Malcolm, 59–60
Classical conditioning, 201
Cleo (cat), 68
Clicks, 62–63, 66–67, 162, 188
Coalitions, 83
Coast Guard, U.S., 49
"Cocktail Party Problem," 178
"Codas," 66, 69–70, 186, 188
"Cogito, ergo sum," 197–98
Cognition, 76, 77–79, 81–82, 85, 101, 103, 117, 122
Cognitive psychology, 203
Cold War, 23
Collins, Judy, 27
Columbia University, 159

Commensal symbioses, 37
Consciousness, 78, 80, 115, 120, 122, 208, 222–23
Cambridge Declaration of Consciousness, 208
"Hard Problem of Consciousness," 122
Conservation, 27–28
Convergent evolution, 80–81
Cook, James, 138
Cooperation, 4–5, 38–39, 48–49, 51, 70, 165, 203
Copying. *See* Mimicry
Corbyn, Jeremy, 13
Core Whale Listening Station, 189
Cornell University, 145, 177
COVID-19 pandemic, 50, 176, 190
Cows, 29, 37, 81–82, 91, 131, 133, 208
Crows, 121, 203, 224
Crying, 59, 149
Cultural transmission, 93
Cuttlefish, 91
Cyanobacteria, 38

Dalston Kingsland, 11
Damon, Matt, 52
Darwin, Charles, 150, 200
Data (databases). *See* Big data
David Frost Show (TV show), 27
Davidson, George, 48, *49*
Davidson, Jack, 47
Da Vinci, Leonardo, 87
Death and mourning, 204–5
Decoding animal communication, 18, 167–93, 227. *See also* Artificial intelligence
Cetacean Translation Initiative, 184–91, *187*
Friedlaender's research, 179–84
Raskin and Selvitelle's research, 169–74, 176–78, 183, 190, 226
unsupervised machine translation, 162, 174–75
DeepMind, 159, 227
Deep neural networks (DNNs), 171–73
Deep Voices (The Second Whale Record) (recording), 87–87
Deer, 81–82
Deforestation, 131
Densely Connected Neural Network, 154–55
Descartes, René, 197–98, 199, 201
"Design features" of language, 93, 98
De Waal, Frans, 90, 204–5
Discreteness, 93
Displacement, 93
DNA (deoxyribonucleic acid), 17, 30, 223–24
Dogs, 29, 50, 56, 112
Dolphins, 3, 15
anatomy of, 62–63
brains of, 73, *73,* 77–78, 82–83, *84*
classification of, 42–44, 160–61
death and grief, *205,* 205–6
Goodwin and, 141–42
human symbioses, 49–52, 85–86
Mirror Self Recognition test, 119–22
Payne and, 21–22, 26
stampedes, 161, 178
strandings, 50–51, 60–61, 69
Wild Dolphin Project, 183–84
Dolphin communication, 66–67, 85–86, 110–27
bubble rings, *112,* 112–13
Herman's experiments, 117–18
Lilly's experiments, 114–16
Oswald's research, 161–63, 171, 222
Reiss's experiments, 112–15, 116–17, 119, *120,* 120–21, 123, 125–27
signature whistles, 113–14, 161–63, *164*
Dolphinville, 206–7
Dolphin worship, 206–7
Dominican Republic, 217–18
Drake, Francis, 164
Drones, 17, 180–81, 186

Druyan, Ann, 31–32
Dylan, Bob, 19, 27

EARS (Ecological Acoustic Recorders), 139–40
Earth, future of, 32
Earth Day, 28
Earth Species Project (ESP), 176–78, 183, 191
Echolocation, 14, 62–63, 66–67, 68–69, 81, 111
"Ecoacoustics," 149–50
Edison, Thomas, 145
Electric eels, 91
Elephants, 17, 170, 203
  brains of, 77, 80
  menopause, 46
  Mirror Self Recognition test, 121
  mourning, 205
  sound and communication, 91, 97, 147, 149
Elephant seals, 2
Eliot, T. S., 212
Eliza (boat), 59
Emiliano (macaque), *95,* 95–96
Endangered species, 43–44, 131, 136
Engesser, Sabrina, 107
English Channel, 56
Enlightenment, 197–98
"Escorts," 215
Ethology, 201–3
Europa, *140,* 141–44
European Food Safety Authority, 208
Evolution, 43–45, 51, 55, 70, 80–81, 82–83, 136, 195
Exceptionalism, 196, 201, 204, 207–8, 210
Extinction, 29–31, 130–31, 210–11, 228

Factory ships, 30
Fahey, Eugene, 209
False killer whales, 50, 77, 121
"Farewell to Tarwathie" (song), 27
FathomNet, 158–59
Fearey, Jack, 161, 177–78
Ferdinand and Isabella of Spain, 200
Feynman, Richard P., 71
Figurative patterns, 136–37
Fin whales, 3, 30, 43
Firth, J. R., 172–73
Fitch, William Tecumseh Sherman, III, 94–96
Fleas, xv–xvi, xviii
Flipper (TV show), 28
Flipse, Gene, 212–12, 214, 215, 219–20, 221–22
Fluking, 8
"Footprints," 214
Foraging, 39, 202
Forebrain, 77, 96
Fournet, Michelle, 176–77
Fouts, Roger, 116
Fox-Case Movietone Corporation, 145
Frediani, Jodi, 219, 221
Free will, 122
Friedlaender, Ari, 179–84, *180,* 210, 217
"Friendly whales," 36
Fruit bats, 158, 177, 178
Fukuzawa, Yukio, 152
Fungus, 38

Gabriel, Peter, 184
Genetics, 195, 223–25
Genome, 223–25
George (snail), 130–31
Gero, Shane, 186
Gershenfeld, Neil, 184
Gibbons, 121, 147, 175
Gibson, William, 208
Gift exchange, 184
Ginsberg, Allen, 115
Goats, 29, 41–42
Goby fish, 37–38
Golden Gate Bridge, 124
Gomero, Silbo, 177
Goodall, Jane, 191

Good Morning America (TV show), 11
Goodwin, Beth, *140,* 141–44, 147
Google, 154, 157, 159, 172, 184
Google Research, 185
Google Translate, 18, 170–71, 174
Gorillas
  brains of, 77, 80
  communication, 94–95, 98–100
Gove, Michael, 208–9
Grammar, 89, 93, 171
Gray whales, 3, 13–14, 43, 51
Great Chain of Being, 198–200, 207
Greenpeace, 28
Gregg, Justin, 112–13, 116, 118
Grief, 205–6
Gruber, David, *185,* 185–86, 189
Guinee, Linda, 25
Gyroscopes, xvii, 16, 170

Hakalau Field Station, 133–34
Hakalau Forest National Wildlife Refuge, 132–35
Happywhale, 153–58, 162, 165, 171–72, 192, *192*
"Hard Problem of Consciousness," 122
Hart, Patrick, 132–38, *134*
Harvard University, 185
Hawaiian Islands, 128–44
  extinctions, 130–31
  Hart's research, 132–38
  Lammers's reseach, 139–41
  Listening Observatory for Hawaiian Ecosystems, 131–35, 139
Hawaiian tree snails, 130–31
Hearing, 21, 24, 60, 63–64
"Heat runs," 4
Henry VIII of England, 164
Herman, Louis, 117–18, 122–23
Herzing, Denise, 183–84
Hispaniola, 218
History of Animals (Aristotle), 199
Hockett, Charles, 92–93, 98, 99, 222
Hof, Patrick, 72–75, *75,* 76–80, 83, 86, 222
Hominids, 194–96
Homo sapiens, 194, 195, 223–24
Honey badgers, 37, 38–39
Honeybees, 91, *92,* 93, 202
Honeyguide birds, 37, 38–39
Hooke, Robert, xv–xvi
Hoover (seal), 97
Horses, 39, 78, 81–82
Howe, Margaret, *115*
Hubble Deep Field, *168,* 168–69
Hubble Space Telescope, 167–68, 176, 190
Human-animal symbioses, 39–42
  Jumper and Jack, 40–41, *41,* 42
  Namaqua people of Namibia, 41–42
Human brain, 76–78, 81, *84*
  communication and, 96–97
Human-cetacean symbioses, 42–43, 45–52
  bottlenose dolphins of Brazil, 49–50
  Killers of Eden, 45–49, *47, 49,* 51
Human Genome Project (HGP), 223–25
Human language, 92–93, 176
Humans, Animals, and Robots (conference), 94–95
Hummingbirds, 91
Humpback dolphins, 50–51
Humpback whales, *34, 129,* 183
  aggression of, 36, 51
  author's breaching incident. *See* Breaching incident of author
  author's observations, 1–9, 35–36, 51
  author's Silver Bank dive, 212–22, *226*
  classification of, 43
  communication. *See* Whale communication
  evolution of, 55
  feeding, 4–5, 43–44, *44,* 45
  Happywhale, 153–58, 162, 165, 171–72, 192, *192*
  interspecies cooperation, 51

Payne's recordings, 22–29, 147
population, 25, 30, 31
recordings, 124, 139–44, 183
Reiss's playback experiment, 123–25
sex differences in song, 151
size of, 3, 5–6
tagging, 179–84, *180, 181,* 210
vocalization. *See* Whale song
Humphrey (whale), 123–25
Huygens, Christiaan, 167
Hvaldimir (whale), *193*
Hydrophones, 23, 140, 141, 147, 186. *See also* Sound recording devices
Hyphae, 38

Ibn Bājja (Avempace), 199
Icahn School of Medicine at Mount Sinai, 11, 72–73, 87
'I'iwi birds, 132, 135, 137, *139*
Imitation, 97–98, 100, 109, 113–14
Imperial College, 185
"Imprinting," *202*
Inner ear, 71
Inside Nature's Giants (documentary), 56–68
Insula, 81
Intellect, 201
Intelligence, 78–79, 83–84
defined, 76–77
International Bioacoustics Society (IBAC), 147–51, 160–61, 163
International Monetary Fund (IMF), 210
International Whaling Commission (IWC), 28
Interspecies Internet, 184
IPhones, xvii, 15
Ivory-billed woodpeckers, 145

Jack (baboon), 40–41, *41,* 42, 102
Japanese whaling, 28
Jeffries, Michelle, 98
Jet Propulsion Laboratory, xvii
Julius Caesar, 56
Jumping spiders, 149
Jungle fowl, 105
Jupiter Research Foundation (JRF), 140–44

Kamehameha I, 144
Kangaroos, 200
Kanzi (bonobo), 98–99, *99*
Katungal, 45–49
Kelley, William H., 59
Kellogg, Peter Paul, *146*
Kestrels, 136
Kētōs, 42
Kickstarter, 191
Killers of Eden, 45–49, *47, 49,* 51
Killer whales, 2, 3, 44, 50, 122, 206
brain size, 77
death and grief, 205–6
humpbacks and, 36, 51, 153–54
mimicry, 113
Kimmerer, Robin Wall, 35
Kingsolver, Barbara, 19
Kinloch, Charlotte, 1–2, 5–9, 13, 18, 179
Koko (gorilla), 102, *102*
Kona giant looper moths, 131
Koshik (elephant), 97
Krill, 43, 210
Kuczaj, Stan, 78
Kuhn, J. J., *146*

Lammers, Marc, 139–41
Language, 89–94
definition of, 89–90
natural, 92–93, 99–100, 118, 222
sign, 89, 98–99, 102
use of term, 89–91
Larynx, 96, 97, 149
Laysan honeycreepers, 131
Leaping. *See* Breaching
Leary, Timothy, 115
Leiden University, 150
Lemurs, 80, 105
Leopards, 104

Lichens, 38
Light spectrum, 91
Lilly, John, 114–16, 206
Listening Observatory for Hawaiian Ecosystems (LOHE), 131–35, 139
"Little Gidding" (Eliot), 212
Lobsters, 209
Lofoten Islands, 184
"Logging," 35–36
Long-tailed tits, 119
Lorenz, Konrad, 202, *202*
LSD, 115
Lupas, Andrei, 159
Lyell, Charles, 200

Macaques, 95–96, 121, 177
McCarthy, Cormac, 20
Macdonald, Helen, 194
Macfarlane, Robert, 220–21
Machine learning, 137–38, 144, 154–60, 184, 213–14
McVay, Scott, 24–29
Magnetic Resonance Imaging (MRIs), 74
Magpies, 198
Mammograms, 214
Manhattan Project, xvii, 225
Manta rays, 14, 158
Marine ecosystems and role of whales, 210
MarineLand Ontario, 111
Marine Mammal Center, 123–24
Marine Mammal Protection Act of 1972, 28
Marino, Lori, 82–83, 206
Martian, The (movie), 52
Mass strandings, 50–51, 60–61, 69
Matrilineal, 46
Mauna Kea, 129–30
Mauritius, 136–37
Max Planck Institute of Developmental Biology, 159
May Maru (boat), 142
MDMA, 149
Melville, Hermann, 30, 181
Menopause, 46, 203
Merlin (app), 191
Mice, 91, 158, *158*
Microscopes, xv–xvi
Migration, 4, 141, 143, 191
Mikolov, Tomas, 172–73
Mimicry, 94, 97–98, 113–14
Minke whales, 3, 43, 71–75, 76–77, 79–80
Mirror Self Recognition (MSR) test, 119–22
MIT (Massachusetts Institute of Technology), 184, 185, *189*
Moko (dolphin), 51
Molyneux, Emery, 164, 196
"Monkey lips," *62,* 64–65, 66
Montaigne, Michel de, 201
Monterey Bay Aquarium Research Institute, 158–59, 170
Monterey Bay National Marine Sanctuary, 1–9, 31, 51, 170
  author's breaching incident. *See* Breaching incident of author
  author's whale observations, 1–9, 35–36
Morality, 90, 204
More, Henry, 197
Moss Landing, 1–2, 8
Moths, 91, 131
Mountains of the Mind (Macfarlane), 220
Mount Everest, 130
Mourning, 204–5
Mutualistic symbioses, 37–40, 42–50, 51
  Jumper and Jack, 40–41, *41,* 42
  Killers of Eden, 45–49, *47, 49,* 51

Namaqua people, 41–42
National Aquarium (Baltimore), 113
National Geographic, 28
National Geographic Society, 185
Natural language, 92–93, 99–100, 118, 222

Natural philosophy, 200–201
Natural selection, 80–81
Nature (journal), 159
Navy, U.S., 23, 49, 98, 165
  sonar, 60–61, 69
Neanderthals, 194–96, 211
Nēnē, 133–34
Neocortex, 77, 81
Neural networks, 154–55, 162, 171–73
Neurons, 76–77, 80–82
New York Aquarium, 113
New York Supreme Court, 209
New York Zoological Society, 22
Nimoy, Leonard, 111
Nobel Prize, 202
Noc (whale), 97–98
Nonhuman Rights Project, 209
Northern Arizona University, 103

Ocean, Joan, 206–7
Octopuses, 203, 208, 209
'Ōhi'a trees, 133
Old Tom (whale), 46–49, 49, *49,* 102
Open source, 177
Operant conditioning, 39–40, 42
Orangutans, 84, 94, 98, 121
Orcas, 14, 17, 206–7. *See also* Killer whales
  mimicry, 113–14
  play, 112
  population, 49
  use of term, 44
Osborne, George, 13
Oswald, Julie, 161–63, 222
Otten, Alice, 48
Otters, 1, 177
Outing (magazine), 59
Owls, 21, 131
Oxford University, 177

Papahānaumoku, 129–30
Parakeets, 136
Parasitic symbioses, 37
Park, Jinmo, 154–55, 171–72, 222
Parrots, 39–40, 203
Pattern recognition, xvii–xviii, 18, 66, 136–37, 144, 172–73, 213–14, 227
Pavlov, Ivan, 201
Payne, Katy, 23, 25–26, 31
Payne, Roger, 19–29, *21,* 31, 97, 207–8
  Cetacean Translation Initiative, 184–85, 190
  whale recordings, 22–29, 31–33, 87–88
Peduncle, 7, 153–54
Pegwell Bay, 56–68, *62, 65, 67*
Penis, 62, 67
Pepperberg, Irene, *100,* 100–102, 108, 116
Pepys, Samuel, xvi
Peter (dolphin), *115*
Petworth House, 163–64
Phoenix (dolphin), 117–18
Phonographs, 145
PictureThis, 191
Pied babblers, 107
Pigs, 81–82, *84,* 208
Pilot whales, 50, 69, 175
Pistol shrimp, 37–38
Plants, Larry, *10,* 10–11, *11,* 153
Plato, 198–99
Play, 12, 111–12, 114, 201, 206
Pocock, 37
Pointing, 112–13
Polar bears, 56, 158
Polar ice castles and palaces, 214
Porpoises
  classification of, 42–44
  mass strandings, 60–61
Prairie dogs, 105–7, *106*
Pratchett, Terry, 110
Prevarication, 93
Primates. *See also specific primates*
  brains of, 80, 83, 96, 195
  Mirror Self Recognition test, 121

Primate communication, 94–95, 98–100, 101, 108
Jumper and Jack, 40–41, *41,* 42
Kanzi (bonobo), 98–99, *99*
sign-based systems, 94, 98–100
Productivity, 93
Project Tidal, *160*
Protestant Reformation, 199
Pschera, Alexander, 191–92
Puʻukoholā Heiau, 144
Pygmy blue whales, 17
Pygmy sperm whales, 50–51, 55

Ramari's beaked whale, 17
Raskin, Aza, 169–74, 176–78, 183, 190, 226
Raskin, Jef, 169
Rational thought, 198
Rats, 91, 149, 201
Rattlesnakes, 91
Reason, 197–98
Recordings. *See* Sound recording devices; Whale recordings
Recursion, 93
Reidenberg, Bruce, 72
Reidenberg, Joy, 11–13, 53–55, *62,* 127, 221–22
dissection tools, 54, *54*
sperm whale dissection, 57–59, 61–68, *62, 65, 67*
whale brain scan, 71–75, *75,* 77–78, 79–80, 82–83, 86
Reinforcement, 116–17, 159–60
Reiss, Diana, 109, 110–17, 177, 184
captive dolphin experiments, 112–15, 116–17, 119, *120,* 120–21, 123, 125–27
whale playback experiment, 123–25, *126*
Renaissance, 199
Rendell, Luke, 69
Rhyme, 25
Riebel, Katharina, 150
Rice's whales, 17
Right whales, 43, 87
Rigid-hull inflatable boats (RHIBs), 179–81
Ripper (duck), 97
Risso's dolphins, 3
Romans, ancient, 29, 56
Root-Gutteridge, Holly, 108
Rose, Evangeline, 150
Rossellini, Isabella, 111
Royal Society, xvi, 200
Ryan, John, 170–71, 176–77
Rychen, Jörg, *146,* 184

"Saddle markings," 45–46
Safina, Carl, 207
Sagan, Carl, 31–32
Sahara Desert, 158
San Francisco Bay, 123–24
San Francisco State University, 123
Sapir, Edward, 25–26
Saturn, 167
Sayigh, Laela, 177
Scala natura, 198–200, 207
Science (journal), 24, 27, *27*
Scott, Peter, 134
Scott, Robert Falcon, 134
Sea cucumber-star pearlfish relationship, *36,* 36–37
Seagulls, 203
Sea Hunter, MV, 212
Seal bombs, 123–24
Sea lions, 1, 51, 113, 121, 123–24
Security cameras, xvii
Selection bias, 15
Self-awareness, 119–22
Self-directed behaviors, 119–21
Selvitelle, Britt, 169–71, 173, 174, 176–78, 183, 226
Semanticity, 93, 99
Sense, 38
SETI (the Search For Extra Terrestrial Intelligence), 177

Sex differences, 150–51
Sheep, 29, 81–82, 131, 133, 208
Sheldrake, Merlin, 151
Sherman, William Tecumseh, 94
Siamese fighting fish, 119
Signaling, 38–39, 40, 50, 91, 223
Signature whistles, 113–14, 161–63, *164,* 174–75
Sign language, 89, 98–99, 102
Silicon Valley, 170
Silver Bank, 212–22
Slobodchikoff, Constantine "Con," 103–4, 106
Sloths, 91
Snails, 130–31
Social Dilemma, The (documentary), 169
Social networks, 83, 187
SoFi Soft Robotic Fish, 189
Soft robotic fish, 186, 189, *189*
Sonar, 60–61, 63, 69
Songs of the Humpback Whale (recording), 27–28
Sound recording devices, 145–52
  Goodwin's autonomous sea vehicles, *140,* 141–44, 147
  Hart's forest reserve, 132–38, *134*
  Payne's singing humpback whales, 24–29, 31–33, 87–88
  Reiss's dolphinariums, 123–25
Southerland, Ken, 154–55
Sowerby's beaked whales, *43*
Space Telescope Science Institute, 167
Spectrograms, 24–29, *126,* 145–47
Speech, 94–95, 98, 108–9, 174
Spermaceti, 64–66, *65*
Sperm whales, 59–60
  anatomy and communication, 61–70
  birth, 219–20
  brain scans, 71–75, *75,* 79–80
  Cetacean Translation Initiative, 184–91, *187*
  family groups, 69
  Pegwell Bay dissection, 56–68, *62, 65, 67*
Spinelli (mouse), 72
Spinner dolphins, 206
"Spontaneous generation," xv
Spotted dolphins, 85–86, *86*
Star pearlfish-sea cucumber relationship, 36–37
Star Trek IV: The Voyage Home (movie), 1
Stewart, Patrick, 20
Strandings, 50–51, 60–61, 69
Sublime, 220
Sun, future of, 32
Swallow, George, 97
Symbioses, 36–40, 42, 47, 48. *See also* Mutualistic symbioses
  Killers of Eden, 45–49, *47, 49,* 51
Syntax, 18, 93, 118, 159, 163, 174, 175, 203
Syrinx, 97

Tahlequah (whale), 205–6
Tail fluke, 4, 7–8, 153–54
Tarsiers, 91
TED Audacious Project, 185
Telescopes, xiv
Tencent, xvii
Tetris, 174
"Themes," 24
Thermometers, xvii
Thomas, Ted "Guboo," 51–52
Thorndike, Edward, 201
Thunberg, Greta, 210
Time (magazine), 13
Tinbergen, Nikolaas, 202
Titan, 167
Toothed whales, 43–44
Tufts University, 21
Turing, Alan, 145
Twain, Mark, 226–27
Twitter, 185
"Two-way communication system," 108
Tyrannosaurus rex, 6

University of California, Davis, 203
University of California, Santa Cruz, 182
University of Cambridge, 208
University of Hawai'i News, 138
University of Hilo, 131–32
University of Lincoln, 108
University of Maryland, 150
University of Sussex, 148
University of the Basque Country, 174
University of Zurich, 107
Unsupervised machine translation, 162, 174–75

Van der Gucht, Estel, 80
Van Leeuwenhoek, Antonie, *xiii,* xiii–xvi, 167, 200
  animalcules, xiv–xv, *xv,* 167, 168–69, 225
Vaquita porpoises, 43–44
Ventilation, 213
Ventral pleats, 6
Vergara, Valeria, 177
Vervet monkeys, 104, 105
"Vocal clans," 69–70
Vocalizations, 12, 24, 94–95, 96–98, 174–75. *See also* Alarm calls; Bird vocalization; Whale song
Vocal learning, 96–103, 113–14
Von Economo Neuron (VEN), 80–82
Von Frisch, Karl, 202
Voyager Space Probes, 31–33

Wākea, 129–30
Wallace, Alfred, 200
Washoe (chimpanzee), 99, 116
Watlington, Frank, 23, 24
Wave Gliders, 140–44
Webb, Wesley, 152
Whales. *See also specific species*
  author's breaching incident. *See* Breaching incident of author
  author's film shoot, 15–18
  author's sighting, 1–9, 35–36
  breaching, 5, 7, 12–13, 217
  classification of, 42–45, 160–61
  death and grief, 205–6
  evolution of, 6, 43–45, 55
  exinction, 30–31, 210–11
  family groups, 69
  feeding, 4–5, 43–44, 210
  friendships, 4–5
  mass strandings, 60–61
  Reidenberg's dissection, 53–60
Whale barnacles, 6–7, 9, 12, 37, 53, 153–54
Whale brain, 71–77, 80–85, 165
  evolution of, 82–83
  scans, 71–75, *73, 75,* 76–77, 79–80
Whale communication, 12–14, 17, 83, 97–98. *See also* Whale song
  anatomy and production of sound, 61–68
  Cetacean Translation Initiative, 184–91, *187*
  clicks and echolocation, 14, 62–63, 66–67, 68–69, 81, 111
  decoding. *See* Decoding animal communication
  sound recording. *See* Whale recordings
Whale data. *See* Big data
Whale detonation, 57, *58*
Whale-human interactions, 13–16, 206–7
Whale oil, 30, 45
"Whale-o-phones," 140–44
Whale recordings, 17–18, 165–66, 179–80
  EARS (Ecological Acoustic Recorders), 139–40, 141
  Payne's recordings, 22–29, 31–33, 87–88
  Reiss's playback experiment, 123–25
  spectrograms, 24–29, *126,* 145–47
  Watlington's recordings, 23, 24
  Wave Gliders and Europa, *140,* 141–44

Whale song, 13–14, 24–26, 83, 170–71, 221–22
Payne's recordings, 24–29, 31–33, 87–88
rhyme, 25
sex differences, 151
"themes," 24
Watlington's recordings, 23, 24
Whale tagging, 179–84, *180, 181,* 210
Cetacean Translation Initiative, 184–91, *187*
Whaling and whalers, 22, 23, 28, 29–31, 45–49, 59–60, 70
"Whisper," *88*
White supremacy, 207
Whole language, 87–88
Wide, James Edwin "Jumper," 40–41, *41,* 42
Wild Dolphin Project, 183–84
WILDME, 158–59
Williams, Bob, 167–69, 176
Wilson, Edward O., 167
Wise, Steven, 209
Word relationships, *172,* 172–73

Yosemite National Park, 78–79
YouTube, 10, 14
Yuin nation, 45–49

Zebra finches, 97, 177